Hope for a planet in crisis

The Forest Underground

Tony Rinaudo

ISCAST Melbourne

© 2021 Tony Rinaudo
Published by ISCAST.
PO Box 40, Forest Hill, Victoria 3131, Australia
editor@iscast.org
www.iscast.org

CHRISTIANS IN SCIENCE AND TECHNOLOGY

ISCAST is the Institute for Christianity in an Age of Science and Technology: a network of people, from students to distinguished academics, exploring the interface of science, technology, and Christian faith.

Scripture quotations marked NIV are from the The Holy Bible, New International Version,® NIV® Copyright © 1973, 1978, 1984, 2011 by Biblica, Inc.™ Used by permission. All rights reserved worldwide.

Scripture quotations marked RSV are from the Revised Standard Version of the Bible, copyright © 1946, 1952, and 1971 the Division of Christian Education of the National Council of the Churches of Christ in the United States of America. Used by permission. All rights reserved.

Apart from any use as permitted under the Copyright Act 1968, no part may be reproduced by any process without prior written permission from the copyright holder. Every attempt has been made to trace and obtain copyright. Should any copyright not be acknowledged or any error be observed, please write to the publisher.

Hardcover: ISBN 978-0-6450671-1-8
Paperback: ISBN 978-0-6450671-2-5

A catalogue record for this work is available from the National Library of Australia.

Proceeds from the sales of this book will support the global FMNR movement.

For years, I have longed to see this biography written. And it lives up to my every hope. Having travelled with Tony I knew he was a magician of forests and a modern miracle worker of environmental regeneration. But because he is softly spoken and so humble it was difficult to fully piece together this extraordinary vision. Until now. From Sicilian family roots, settling in rural Victoria, this is a story of breathtaking global impact. Indeed, it may yet save us all!

Tim Costello AO
CEO World Vision Australia 2004–2016

There are three movements that easily have been the leaders in restoring the developing world's soils and thereby making life better for village people. One is the FMNR movement led by Tony Rinaudo, the second is the conservation agriculture and green manure/cover crop movement headed by Valdemar de Freitas and Ademir Calegari of Brazil (in which I am involved), and the third is the holistic management of animals movement led by Allan Savory. Each of these movements has improved the lives of many millions of farmers. But Tony's FMNR movement has done far more, per dollar spent, to improve the lot of the world's smallholder farmers, than any other major agricultural movement in world history, including the famous Green Revolution. That may sound like it must be a tremendous exaggeration, but I have personally done evaluations of well over 50 development programs on four continents, including programs that form part of each of these three movements, and the numbers easily support what I have said.

Roland Bunch
Founder and CEO Better Soils, Better Life

Tony Rinaudo is a shining example of what perseverance and courage can achieve. Faithfully following his calling, Tony has worked diligently to ensure that some of the world's most vulnerable populations can enjoy food security and live a life of dignity. Little did he know that his concrete and simple method of helping "underground forests" re-green dry areas would improve the lives of millions. As we face an aggravating climate crisis, Tony's example gives hope. I am certain that his visionary work has the potential to change the world.

Ole von Uexkull
Executive Director of Right Livelihood

Tony Rinaudo had an epiphany in Niger in the mid-1980s about the power of Farmer Managed Natural Regeneration (FMNR) to transform the African drylands. FMNR is a practice that directly confronts the conventional paradigm of agriculture: crops ought to be produced in clean, treeless fields.

Such a new paradigm must be justified—by farmers' experience and by science. The evidence base is growing rapidly. There is now solid documentation of millions of smallholders around the world successfully practising FMNR on their farms, community forests, and grazing lands. There is also a growing body of literature validating farmer observations on the multidimensional benefits of this practice. FMNR can provide trees for food, fuel, and sellable goods: this is an obvious way to overcome food insecurity, increase incomes sustainably, and provide these families with dignity.

Tony Rinaudo's epiphany about FMNR in Niger and his steady persistence in building awareness and expanding training efforts across many countries have inspired scale-up efforts that we see taking root all over Africa and across the world.[1]

Dennis Garrity
Drylands Ambassador for the UN Convention,
Senior Fellow, World Agroforestry Centre,
Chair, Global Evergreening Alliance

Dedication

For my dearest Liz,
who shared my vision from the start and
whose love and support have never wavered,
and to my long-suffering children,
Ben, Melissa, Daniel and Sarah,
who tolerated an all-too-often absent father.

Contents

	Foreword	9
1	Paradise	13
2	Roots	23
3	A child's prayer	31
4	The call of the Sahel	39
5	A welcoming smile is better than a welcome mat	57
6	Sowing into the wind	93
7	The forest underground	107
8	May your God help you	121
9	The lizard also drinks from the chicken's water bowl	135
10	Year of the Tankarki	147
11	A global movement	167
12	One billion hectares	193
	Acknowledgements	207
	Notes	211
	Photo credits	217

Foreword

I first met Tony Rinaudo in December 2018 shortly after he received the Right Livelihood Award. I was so impressed that I spontaneously decided to make a film about him. Only a few months later my camera crew and I attended the Beating Famine Conference in Bamako and then spent several weeks travelling with Tony in Mali, Ghana and Niger, places where he has worked for decades. I could see for myself: his method works. Hundreds of thousands of prospering farmers and their families are practising Farmer Managed Natural Regeneration (FMNR). In Niger alone six million hectares of denuded farmland have seen tree density increase from an average of four trees per hectare in 1980 to 40 trees per hectare today.

We soon became allies and friends. I was impressed by the energy and passion Tony put into the Beating Famine Conference and by the devotion of his wife Liz in welcoming guests and delegates. Agronomists and supporters of FMNR from all over the world applauded when Tony praised the value of trees not only for restoring deteriorated soils but, above all, for restoring hope. Indeed, wherever Tony goes, hope is in the air. Now the numerous projects and initiatives of individuals, NGOs and even governments are merging into a real social movement.

What is most striking is Tony Rinaudo's character. This is the root of his success in promoting reforestation and agroforestry in villages. To see the grateful and happy farmers, women and children is overwhelming. Their devotion, faith and passion is tangible. These people's lives have been completely transformed.

Tony's techniques and the efforts of farmers that practise them are modest, but they are successful. If we want to achieve our climate goals, it is time to change perspective. It is time for our farmers, agricultural cooperatives and representatives of industrial agriculture to shed their Western arrogance and learn from these unknown farmers who perform small miracles day after day.

Tony shows us how. During the famine of 1984–1985 he lived with the poorest of all farmers. He experienced penury himself. Ever since, he boils the water he needs the next day in his hotel room. No plastic bottles, no industrialised water for him. No air conditioning either; he just wraps a damp cooling scarf around his neck. To be fit for the day he runs an hour before sunrise, whether in the streets of Kolkata, the rural paths in Bolgatanga, by the muddy slopes of the river Ganges or the dusty Niger. Being a runner myself I joined him on these early excursions under the gaze of amused locals.

One evening, overlooking the Niger river, Tony told me that Africa could easily feed its entire population, and even that of the world. I was, at first, as doubtful as Thomas. But travelling further with him, I learned to understand the almost unlimited opportunities for agriculture on this huge, not-yet overpopulated continent.

Nevertheless, the situation is desperate for the one billion people still living as smallholders around the world. Their yields are shrinking dramatically. Up to 700 million people will be obliged to leave their homelands during the next decades because of the rapid pace of desertification. This is no vague prophecy of doom, but the forecast from over 100 scientists at the Bonn-based Intergovernmental Science-Policy Platform on Biodiversity and Ecosystem Services (IPBES).

In November 2019, my team and I followed Tony to India. His purpose was promoting FMNR through workshops and in-field training in villages in the eastern state of Jharkhand, and later in the huge western state of Maharashtra. To finish the trip, he spent three days in New Delhi meeting with Indian agricultural specialists and officials to evaluate the validity of his method for a population of at least 300 million farmers. This type of lobbying clearly was not up his alley. Yet it had to be done. In the last decade alone, more than 100,000 farmers in India took their own lives because they saw no other way out of their misery.

It became clear to me that, while we are all paralysed by the climate change predictions, an agronomist, a missionary, a simple man from Australia may single-handedly have found the solution. Reforestation of almost all barren land is plausible, at very low cost, thanks to FMNR. Since his work began 30 years ago, 240 million trees have grown in Niger alone. Today his dream is to reforest a billion hectares on our planet by inspiring others. This is the most ambitious yet most affordable proposal to stop rising temperatures.

Good timing! The planet badly needs it. It is no exaggeration to say that Tony Rinaudo may save the planet. His nickname is, rightly, the "Forest Maker." Tony Rinaudo needs and deserves disciples all over the world.

Volker Schlöndorff[2]
Director of the film *The Forest Maker*

CHAPTER ONE
Paradise

Our house in Myrtleford, a small, country town in north-eastern Victoria, sat at the foot of Reform Hill. From the lookout, I could look down on the township: miniature buildings, homes, cars and people going about their business. I could see the confluence of the Buffalo and Ovens Rivers and the rugged cliff faces of Mount Buffalo. The tranquil beauty of the blue hills and the narrow, green Ovens Valley imbued me with a strong sense of place and belonging. Even though I have lived away for longer than I have lived at home, this affinity remains strong to this day.

These hills and valleys provided me with a perfect playground. I belonged to a small band of children who lived at the end of Elgin Street's cul-de-sac. We often played together and kept our guardian angels busy. Occasionally, our mothers would learn of a near miss with a snake, mine shaft or tree climb, and our adventures would be banned until we wore down their resistance and were allowed to roam free again. Cowboys and Indians was a favourite game. When alone, I ran downhill at full pelt. In my dreams I was airborne, crashing through strong silk strands strung between trees by giant spiders.

I am the third child in our family of four boys and two girls. My younger brother Peter and I were inseparable, always bushwalking, fishing, riding bikes—though, I suspect he often came along to please me rather than out of any enthusiasm for the outdoors. After having four sons, Mum dearly wanted a daughter. Dad had lost his only sibling to leukaemia when she was still a young mother of two boys, and so when my sisters Cathy and Josie were born, they were welcomed with much joy. They were special to me too and I loved helping Mum look after them.

Every Sunday morning after church, Dad pulled out his Box Brownie camera. While Mum prepared spaghetti for lunch,

I was born and raised where the forest meets the farm. The Ovens River valley viewed from Mount Buffalo.

Clockwise from Mum: Peter, Sam, me, Cathy and Josie in 1966. Joe was at boarding school.

he arranged a quick portrait of us children, still in our Sunday best, in front of the camelia bush. For the first few years we were taller than the camelia, but eventually it surpassed us in height. After lunch, we all piled into our Ford Falcon station wagon for the 40-minute drive from Myrtleford to Wangaratta to visit Dad's parents, Nanna and Nanu. There were no seat belts in those days. When we boys played up in the back seat Dad would swing round and with his free arm whack those of us who didn't duck quickly enough. Mum was not impressed. Apart from wanting to protect us, Mum was a nervous traveller and worried that the distraction would cause an accident.

Once past the Beechworth turnoff, the country opened up and the broader plain was framed by the treeless Murmungee Hills. Can hills speak? Maybe not in words. Yet in their nakedness, they seemed to be grieving and crying out for help and restoration. As we drove to Wangaratta each Sunday, in my mind's eye I was on those hills toiling in my gumboots, shovel in hand, planting trees and plugging the deeply eroded gullies.

Sometimes we returned home in the dark. At sections of the Great Alpine Road the branches of the huge gum trees on either side of the road met above us. As the car sped through the night, the headlights illuminated the trunks and branches. An enchanted cave appeared in front of us and disappeared into the darkness behind us. Not satisfied with these short tunnels, I mentally filled the gaps by planting the missing trees!

The Jaithmathang people

The trees were silent witnesses to the past. For how many centuries had they sheltered and nurtured the Jaithmathang (Ya-ithma-thang) Aboriginal people? For how many summers

The Murmungee Range from Beechworth Gap. The bare hills and naked gullies disturbed me.

had these trees heard their now-lost Dhuduroa language and watched over their annual pilgrimage to the high country to harvest and feast on Bogong moths? From up to a thousand kilometres away, each summer, millions of moths migrated to congregate in the cool rock crevasses of Australia's Southern Alps. The Jaithmathang would roast the moths on hot ash and eat them. Their high fat and protein content and sweet, nutty flavour made them a delicacy.

During the cooler months the Jaithmathang occupied the lower reaches of the river valleys. Camps were established on the soft soil of the open, flat country where water and food sources were plentiful—a recipe for disaster in the unequal competition and culture clash to come.

Don Watson's *The Bush* (2014) dispels the early settlers' myth of *terra nullius* or "empty land" which was clearly designed to justify a continental-scale land grab. He cites numerous references from explorers and early settlers to a landscape that in its open orderliness and beauty looked like a "gentleman's park," an "English park," a "French park," an "immense park" or "one stupendous park."[3] In *Fire Country* (2020), Victor Steffensen describes pre-colonisation landscapes as "beautiful with plentiful food, medicines and life."[4]

> The trees were managed to stay on the country, to grow old and become the Elders of the landscape, maintaining their gift of providing life and prosperity for every other living thing within their environment. Aboriginal land management ensured that most of the trees lived to be hundreds or even a thousand years old. They populated the country in plenty, drawing and giving goodness to the ground to provide the essentials for a healthy landscape.[5]

In grappling to come to terms with the "sick" state of much of the Australian landscape today, Steffensen also draws attention to the disconnectedness of most people from the land.

The impact of Europeans

Part way up Reform Hill, a stone monument marks the passage of Hume and Hovell during their 700-kilometre expedition from Sydney, New South Wales, to Port Phillip, Victoria, in 1824. In the wake of these explorers, early squatters took up land to graze sheep and cattle. Abandoned gold mines, mullock heaps, and an enormous, now-silent, rock crusher tell of the gold rush which began in the 1850s. Reform Mine was north-eastern Victoria's most productive underground mine, producing more than 21,000 ounces of gold. Was it possible that even the Kelly Gang, the notorious bushrangers, passed this way while on the run? If we were lucky, we would see a kangaroo or an echidna, and occasionally a venomous brown or tiger snake, but more commonly, we saw the rabbits introduced by European settlers.

The countryside around Myrtleford saw various economic activities ebb and flow in importance over the years—beef and dairy cattle, sheep, flax (during and after World War II), pine, hops, wine grapes, blueberries, olives, walnuts and chestnuts. Tobacco was the drawcard that brought many Italian migrants, including my grandfather, Giuseppe "Joe" Rinaudo, to the Ovens Valley.

From the late 1920s onwards, exotic pine plantations began to replace native vegetation on many of the hills in the district. Indigenous bushland was bulldozed. Thousands of trees were heaped into windrows and burnt. The wood wasn't even used! Steep hills were stripped of all vegetation leaving ground bare for long periods of time. Then the hills were planted with a monoculture of *radiata* pine, native to the Central Coast of California. Walking through these dark, silent forests with no undergrowth was like walking through a dull desert. The only birds were those flying overhead to another destination. I bore a sense of loss. Even as a child this approach seemed very short-sighted and destructive. I did not hate exotic trees, but I was indignant at the enormous waste and disregard for what was already there.

The valley of the Ovens River as observed in 1866 by Austrian-born artist Eugène von Guérard.

Not even the hilltops or valleys were spared as safe havens for indigenous wildlife and vegetation. In the fertile valleys, pesticide spray targeting tobacco crops drifted into the cold, crystal-clear mountain streams in which I loved to fish and swim with my siblings and friends. These streams provided the townships with their drinking water. For a period when pesticides were sprayed from aeroplanes, serious fish kills occurred and swimmers were confronted with the sickening sight and smell of large trout floating past, belly-up.

These same waterways suffered significant damage from gold mining, from the 1850s to 1955. Gold panning and sluicing gave way to the deployment of giant battleship-like dredges which systematically desecrated the once-living waterways, smothering fertile valleys with tailings of gravel and rock. An already-damaged river system was further degraded during my teen years when logs were removed and riverbeds bulldozed to speed the flow of water as a "solution" to damaging floods—floods caused, no doubt, by land clearing on the hills! This severe disturbance destroyed fish habitat and converted the wild and beautiful mountain streams that I loved into sterile drains designed to move life-giving water from the valleys as quickly as possible.

In 1911, 36 gold dredges in the Ovens Valley destroyed 250 hectares of fertile river flat.

Nothing, it seemed, was sacred. Myrtleford was one of the few towns anywhere in Australia that boasted a stately row of mature palm trees in the main street. The trees were like old friends who were always there to welcome visitors to the town centre. I was shocked to go into town one day only to be confronted by empty space covered over with asphalt. The palms had been removed in a perfunctory manner because

I was proud of the palm trees that greeted visitors to Myrtleford. They were removed because they harboured birds and dropped fronds.

In preparation for planting exotic pine trees, native vegetation on steep hillsides was bulldozed into windrows and burnt. The cleared land was left bare for several years—a practice continued to this day.

they harboured pests and dropped untidy fronds on the road. How and why could anyone do this? Where would it end?

I knew that farming was necessary, but I questioned the wisdom of clearing all the indigenous flora and fauna from the land. Why did it require so much destruction? At university, the insight of my daydreams seemed to be confirmed, not in the formal lectures, but in the pages of *Forest Farming* (1976) by James Sholto Douglas. Douglas wrote about how the integration of trees, crops and livestock brought about a sounder ecological balance and greater productivity of food and other materials for clothing, fuel and shelter. It made perfect sense to me, but it contrasted starkly with the approach of early settlers who imported destructive European farming practices. Colonialists saw it as their duty to tame and "civilise" the bush in order to make it "useful." They cleared the trees and killed the wildlife that competed with livestock and crops. In the process, they also removed the First Nations Peoples. Settlers who had been granted land were actually required by the government to clear it of trees in order to keep it. The attitude of early settlers is well summed up by the popular Australian adage, "If it moves, shoot it; if it doesn't, chop it down!"

This saying may not have been aired in university halls, but the only difference between colonial approaches and "modern" agriculture was the air of respectability given to the latter by scientific and economic rationales. Modern agriculture was built on this flawed foundation laid by the European settlers. It meant blinkered mastery of nature through chemistry and engineering. It is characterised by uniformity of crops and livestock for high and ever-increasing yield, and is driven by a desire for higher profits without reference to the environmental cost—loss of ecosystem function, including soil degradation and biodiversity loss. Farmers are under enormous pressure. They have to make a profit to make a decent living and stay viable. But doing so by degrading the land will only put themselves and future generations at peril. Fortunately today, regenerative agriculture, an umbrella term for a host of practices more in tune with nature, is gaining momentum globally.

CHAPTER TWO
Roots

My father's family

To build a better life for his family, Giuseppe Rinaudo migrated to Australia from Sicily in 1926. He had no intention of ever returning to Italy. He planned for his wife Catarina, daughter Domenica, and son Gaetano (Tom, my father) to join him within a year. However, the Great Depression made jobs scarce and it would be seven years before he would see them again. First, he had to repay his own passage to Australia. Then he had to save enough money to bring his family. He took work wherever he could find it. This included labouring on farms and the construction of the Hume Weir on the Murray River.

When I was a boy, I sometimes stayed at Nanu and Nanna's place. After dinner, Nanu would light up a cigarette and talk. I didn't like the smoke but I did like his stories. He told me that, at my age, back in Sicily, he used to transport wine in barrels on a donkey cart for the family tavern in Ramacca and deliver goods for others along the route. One day he fell asleep and was suddenly woken by a robber. He burst into tears. Fortunately, the robber took pity on him. From that day on he took up smoking to stay awake. When I was unhappy, he was unsympathetic. "You do not have to work to support your family. You have been given everything imaginable. And yet, your generation is unhappy. Too many are using drugs, and some are taking their own lives. Why?" He told me of his experience as a swagman—the Australian version of a hobo—walking from town to town looking for work.

> I watched my father earn the trust and respect of clients. He asked questions and listened before speaking.
> He endeavoured to understand what people needed and what values drove their decision making.

After walking all day, I would light a fire under a bridge and sleep there the night. Other swagmen who were total strangers would join me, and we shared what we had, enjoying each other's company. Late into the night we gave each other instructions on places to avoid and where to find work or food. We told tall tales and joked to pass the time. We had nothing, life was hard, but we were happy.

Many migrants got established by farming tobacco. Cultivating and processing tobacco was labour-intensive.

Over the years that he lived on the road, he told me that he only ever saw one case of suicide. A man had hung himself under a bridge where Nanu slept.

By the time Nanna arrived in Australia in 1933, Nanu and his brother Antonio were growing tobacco as share farmers at Whorouly East in the Ovens Valley. Tobacco farming was labour-intensive and farm owners entered into a contract with farm labourers. The farm owner provided the land, equipment and accommodation; the share farmer grew, harvested and sold the crop. After the sale, the profits were shared.

At her reception party Nanna was shocked at the platters of pork, beef and chicken. Being used to living frugally in Sicily and eating meat only twice a year—for Christmas and Easter—she scolded Nanu on his extravagance!

Nanna could knit and crochet intricate lace work. She made beautiful cardigans, place mats, shawls and doilies. This skill had come at a price; her stepmother had been a harsh taskmaster, pinching her fingers every time she made a mistake. Her mother had died when she was young. During the lonely years of waiting to come to Australia she crocheted four double-bed-sized lace covers and suffered the taunt, from her own father, that Giuseppi had abandoned her. She was very self-conscious about her low level of education and had been deeply hurt by the discrimination she experienced when she arrived in Australia. It cut deeply to be ridiculed for her accent and to hear stories of how her husband had often been refused work because he was a "wog." Nanna's beautiful accented lilt still rings in my ears: "An-do-ny, get-a-job-a-wid-a-suit-na-tie!" I could understand why she wanted that for me, but an office job was the last thing I wanted. Being poor does not stop people from being proud or aspiring to a better life. I often ponder the wisdom and values in my grandparents' stories.

My father, Gaetano (Tom) Rinaudo, was born in Ramacca, Sicily. He was eight years old when he arrived in Australia. Each morning, he milked eight cows, separated the cream, fed the calves and walked five miles (eight kilometres) to the local school. Eager to get to know his father, he often followed him into the tobacco fields. However, once he had to be hospitalised with sunstroke after spending too much time in the hot sun. Working long, hard hours on the tobacco farm made him realise this was not what he wanted to do. To encourage him to continue his education, Nanu showed my father an ancient and massive river red gum in the centre of an uncleared block of land. "I am going to chop down all of the smaller trees. This one is reserved for you if you do not finish school." Red gums have extremely hard wood, and large specimens can have a girth of four metres or more. It would not have been easy to cut one down with an axe!

Felling a mature river red gum with an axe would be a daunting prospect.

Dad was inventive, inquisitive and hardworking. After studying motor mechanics and completing his apprenticeship, he went into business with a partner in Wangaratta, selling and servicing Vanguard cars and Massey Ferguson tractors. In 1952, he established Buffalo Farm Equipment in Myrtleford, selling International Harvester trucks and tractors, machinery and supplies. He targeted the booming tobacco industry, but also serviced farmers in the

Dad was enthusiastic and generous. He always supported his children's interests and passions.

surrounding valleys and wider district. He became well known for his fairness and honesty.

My father keenly felt his parents' pressure to succeed. As was common in migrant families, Nanna and Nanu pushed him to be a man and to make something of his life. Naturally they wanted something better for their son. While Dad was devoted to his parents and loved and cared for his family, he was dedicated to his successful business and to his customers.

Dad was often out on calls repairing broken machinery. He helped his customers where and when he could. Many of them became good friends. Unlike my three brothers, I don't have a mechanical bone in my body, but Dad often took me along for the ride. I loved going with him to visit the farms. While he was busy fixing machinery, I went fishing, raided fruit trees or just enjoyed being on the land. While Dad would have loved for me to eventually join the family business, my passion lay elsewhere. Two of my brothers, Sam and Peter, took over when Dad retired in 1990. In 2003 Peter left to set up his own engineering company, Mechanism.

At the garage, Dad had to manage the complexities of his thriving business, including the occasional difficult and demanding customer. But on the road he was a different person. Out on those calls he was doing what he loved and what he was very good at—meeting people and fixing machinery. Whenever he drove through the countryside, he whistled popular Italian tunes: *O Sole Mio*, *Arrivederci Roma*, or *That's Amore*. He loved being with others. Although there were many demands on his time, after the job was done, I don't recall him ever refusing a cup of Italian espresso coffee and a yarn. While I couldn't follow the conversation when it was spoken in Italian, I could see that Dad was in his element and that people respected him. We invariably returned home laden with home-grown produce from grateful farmers. These experiences served me well when I worked closely with thousands of farm families in Niger. The ability to make friends, build trust, empathise, ask questions, listen, and understand others' needs and what drives their decision making are worth more than all the world's development theories combined. Years later, I was surprised—yet not surprised—when a retired sales representative who knew my dad told me, "Your mannerisms are just like your father's."

In what little spare time he had, Dad loved gardening. On Saturday nights, the family watched *Sow What* with Kevin Heinze on ABC television. Dad was very practical; when he saw the program on greenhouses, he built his own from scratch using timber from packing crates. Like him, I found deep satisfaction in growing things. I spent many hours in our large veggie patch helping, watching and learning.

My mother's family

My mother, Carolina Rando, was born in Scari, on the active volcano and island of Stromboli. Dad loved to tell us that Stromboli was so small, when the tide came in the Strombolians would have to swim to dry ground on Sicily! My mother's grandfather, Gaetano Russo, appears in several scenes in Roberto Rossellini's 1950 film, *Stromboli*, starring Ingrid Bergman. In the movie the monotony and conformity of island

My maternal great grandfather appears several times in Roberto Rossellini's 1950 film, *Stromboli*.

life is interrupted by the annual tuna hunt. The sleepy village becomes a hive of activity and model of mutual dependence. As the row boats encircle the catch and draw the nets tight around the trapped tuna, their thrashing brings the calm sea to a boil and the men gaff and pull the giant fish onto their boats.

Life was hard in Stromboli. When boys reached the age of 12 or 14 they were sent to the Americas, Canada, Australia or New Zealand to work for relatives or neighbours who had

Dad loved to socialise. Mum loved to dance. Together they cultivated a wide circle of friendships.

preceded them. In 1904, at age 10, my maternal grandfather, Salvatore (Sam) Rando was sent to Wellington, New Zealand to work with his fishmonger father, Giacomo Rando. Six years later, he set off to New York to work in an uncle's fruit shop. Resenting unpaid employment, he quit and worked as a milkman. He told me that it was often so cold on milk runs he lined his jacket with newspapers. He became Reid Ice Cream company manager in under eight years. From 1917 to 1919 he was a sailor in the US Naval Coast Defence Reserve on the *USS Arizona*. He loved the sea and the navy. Years later, when he read of the sinking of the *USS Arizona* and its 1,177 crew in Pearl Harbour, he wept. A fortuitous meeting with a member of the Vanderbilt dynasty developed into a friendship and the realisation of his dream—a mixed business on Coney Island which allowed him to spend six months in the United States during the peak holiday period and six months in Stromboli. During these visits to Stromboli, he married and had three children, including my mother, Carolina Rando. He was all set to take his wife and three children back to New York with him, but his father, who had by then settled in Melbourne, Australia, with the rest of their family, pleaded with him to join them. Mum was seven years old when they arrived in Melbourne in 1932.

Mum was quiet, reflective and a devout Catholic. She was baptised and confirmed at the Catholic Chiesa di San Vincenzo in Scari. The church played a prominent role in the life of the islanders, providing spiritual nurture and physical protection from the volcanic debris which from time to time rained down on their side of the island. Mount Stromboli has been in almost continuous eruption for the past 2,000–5,000 years. Mum still remembers fleeing to the church when the volcano erupted on 11 September 1930. This is the most violent and

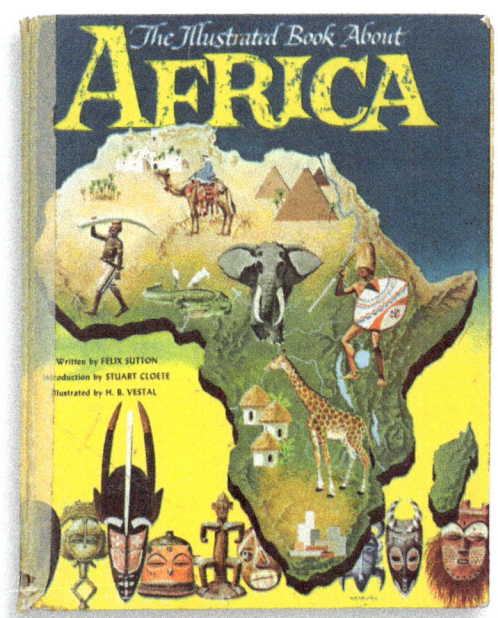

I still have the book about Africa my mother gave me when I was in primary school.

destructive event in the record of Stromboli's activity. The eruption lasted less than one day but caused considerable damage and several deaths.

Mum was a gentle person but when one of us raced through the house screaming a warning—"Quick! Mum's coming! And she's got the wooden spoon!"—we knew we'd pushed her too far. I did not have the emotional intelligence to appreciate her strong and gentle spirit. I was stubborn and very hard on her at times. One day, when she returned from confession, I did something to upset her. "Oh Anthony! You've made me so angry! Now I have to go back to confession!"

Mum complemented Dad, whose Catholic faith guided his principled but perhaps more pragmatic approach to life. Mum's serenity, despite long days dealing with six children, spoke volumes about her understanding of the character of God. It also gave me a framework for understanding the meaning of life. I never questioned whether there is a God who loves and cares for us and that we are put on earth for a purpose. I grew up with a sense that we are responsible for the welfare of those less fortunate than ourselves, and we have a duty to care for the earth and all that is in it.

The first book Mum gave me was a children's picture Bible, which reinforced the Bible stories I heard at church and in school. The second was a book I chose at the local newsagent—perhaps as a reward for good behaviour. It was a picture book about Africa and its history, cultures, peoples, animals and landscapes: desert, savannah, jungle. More than any other continent, I was drawn to the diversity, mystery and wildness of Africa.

CHAPTER THREE
A child's prayer

At St Mary's Primary School in Myrtleford I discovered a passion for reading about other countries. Above all other subjects, I was most enthusiastic about geography. My class teacher, Mrs Weston, recognised this and encouraged me, and Dad always encouraged his children's interests. To display my maps, he mounted sheets of plywood onto my bedroom wall with hinges. By successively swinging each sheet, I revealed another continent—Africa, Europe, South America and so on. From my bed, I could look at the world and dream.

In 1969, at age 12, I was outraged to read that the Cuyahoga River in Ohio, USA, was so polluted it caught fire.[6] Surely, this catastrophe would make world leaders pause and question their reckless pursuit of wealth and development without considering the environment—the scaffolding upon which all life on earth depends. But it didn't. This shook what little confidence I had in the ability of government or industry to make wise decisions about environmental issues. Deforestation in the tropics, oil spills on the oceans and industrial pollution were in the news weekly. In the years following, these regular reports of environmental harm impacted me deeply. I saw denuded hills and poisoned rivers in my own back yard, and an all-out war on nature globally. I imagined an apocalyptic, nature-less future. Leaving aside its exquisite beauty, how could rational adults do this to the living planet that we all depend on for air, water, food and inner well-being? Similar questions were to haunt me in Africa. How could clever, hard-working farmers destroy the trees that made farming possible in the first place and jeopardise their families' futures?

The unbridled destruction made me angry but it was the normalisation and acceptance of violence against nature, the blind belief that so-called progress justified the destruction and loss of natural processes, that caused my strongest reaction.

When a river catches fire, things are bad. The Cuyahoga River in Ohio, USA, caught fire for the fourteenth time in 1969.

All too often the evening news told stories of famine in some war-torn or drought-stricken country. It seemed unfair to me that children just like me were going to bed hungry, through no fault of their own. Why did we use productive land and precious water in north-eastern Victoria to grow tobacco, a toxic weed, while people elsewhere were lacking food and dying of hunger? I could appreciate that issues are never completely black and white, and farmers were simply making a living. Many tobacco farmers, including my grandparents, came to Australia for a better life, fleeing fascism, poverty and a lack of opportunity. I was a beneficiary of their struggle and courage, and, ironically, of the tobacco industry. But I couldn't justify the cultivation of a weed that made people sick while people elsewhere were dying of hunger.

Champagnat College, Wangaratta

After completing primary school, I was sent to Champagnat College, the Catholic Marist Brothers boarding school for boys, in Wangaratta. This was challenging. I was very quiet at first but eventually made good friends. While I had no talent for ball sports, I was a reasonable runner. I always managed to get onto the athletics team and each year we competed against other secondary schools on the rubberised Olympic track in the Victorian capital, Melbourne.

One of our teachers, Brother Gordon, was developing a Murray Grey cattle herd on the school property. I loved helping this gentle giant and learnt how to drive a tractor, erect fences, attend to calving heifers and plant trees. I even managed to convince Dad to buy a cow for me which we agisted on a friend's farm. It was to be the foundation of my future herd. However, the cow turned out to be a wild and stubborn reject and betrayed my poor prospects as a livestock farmer.

Outside of the football season, weekends were long and boring, so I asked permission to form a fishing club with half a dozen other boarders. Brother Gordon supported my idea. Sharing a brew of Brother Gordon's billy tea was the perfect way to cap off an afternoon's fishing. Take one kerosene tin (the "billy"). Fill it with water and place on a raging open fire until the water boils furiously. Throw in two fistfuls of tea leaves and five of sugar. (Brother Gordon's deeply creased hands were as big as dinner plates!) Pour in milk. Break off a small branch from the nearest eucalyptus tree, whack it on the nearest post to knock off the dust and

As a teenager, I helped my brother Joe on his chestnut and berry farm. I loved being outdoors and doing hard physical work.

insects, then stir the tea vigorously. I don't know if it was Brother Gordon's recipe, the camaraderie or the feeling of everything being right with the world when in the bush with friends, but it was the best tea I'd ever tasted. Of course, on hot days, when we were on our own and the fish weren't biting, we boys took the opportunity to shed our gear for a bit of skinny dipping. Thankfully, the brothers never found out!

While a few of the Marist brothers were social misfits, the majority loved God and loved the boys they taught. Their exemplary life of service and self-sacrifice made a strong impression on me and for a while I seriously considered becoming a member of their order.

One year, our class was taken on a retreat to the Mornington Peninsula. It was there that I experienced what in hindsight I would call a conversion experience. The camp director explained how we put on masks and try to be somebody we are not in order to be accepted. But because God loves and accepts us just as we are, we do not need the approval of others. For a teenager struggling to find his place in the world and to be accepted by the "in" group, this was a powerful realisation. It is one thing to assume the faith of your parents. It's quite another to believe and trust in the living God because of your own personal experience and convictions.

Richard St Barbe Baker: Man of the trees

One day I accompanied Dad to visit family friends Greg and Carmen Saric. As a young man Greg fled communist Yugoslavia, travelling alone and on foot. To avoid detection, he slept in hiding by day and walked by night, often stumbling and falling into thorn bushes. Eventually Greg arrived in Australia, where he met and married Carmen, his Filipina wife. Mum and Dad

Until his death at the age of 93, St Barbe Baker impressed upon presidents, kings, technocrats and tribesmen alike the need to care for trees.

and other members of the Myrtleford Rotary Club helped Greg and Carmen settle in the district. As we walked through his otherwise-empty tobacco-sorting shed, we passed a trailer load of old library books which had been unceremoniously dumped on the floor. In that large pile of books, two dull-green, nondescript volumes caught my eye: *I Planted Trees* and *Sahara Conquest*, by Richard St Barbe Baker. St Barbe Baker was from Hampshire in the United Kingdom and travelled the world to campaign for the preservation of forests and the sustainable management and utilisation of trees. He absorbed the lessons of history and used what he learned to look into the future. He warned what would befall all life on earth if we continued on our foolish trajectory of unfettered destruction of forests. Until his death at the age of 93, St Barbe Baker was a man of action, driven by his love of trees and his deep faith to inspire and mobilise others. Through his audacity, eccentricity and sheer good fortune, he impressed upon presidents, kings, technocrats and tribesmen alike the need to care for trees. And all this he did in faith and on a minimal, often meagre, budget.

I read his books from cover to cover. I was comforted that there was at least one adult who cared about nature, and I felt challenged by his example. I learnt that deforestation was not an inevitable consequence of progress. St Barbe Baker mobilised volunteer armies of tree planters around the globe and demonstrated that something could be done to combat the destruction while delivering a balanced approach to conservation and production. St Barbe Baker wrote:

> We had better be without gold than without timber. Wood is necessary to civilised life, and therefore it is a basis of civilization. The greatest value of trees is probably their

beneficent effect upon life, health, climate, soil, rainfall and streams. Trees beautify the country, provide shade for humans and stock, shelter crops from wind and storm and retain the water in the soil at a level at which it can be used by man. The neglect of forestry in the past has accounted for the deserts that exist, because of the fact that when the tree covering disappears from the earth, the water-level sinks ... When the forests go, the waters go, the fish and game go, crops go, herds and flocks go, fertility departs. Then the age-old phantoms appear stealthily, one after another—flood, drought, fire, famine, pestilence.[7]

I was mindful of St Barbe Baker's example as I approached the end of my schooling and the possibility of going to university. I thought that the best way I could be of use was to study agriculture. I loved growing things, and this would provide opportunities to work directly with poor farmers. Could I help people to farm with methods that were both nature-friendly and productive?

These issues weighed heavily on me and I longed to be part of the solution. I felt angry about environmental destruction, injustice and poverty. I felt powerless in the face of the indifference of many adults. I lacked the maturity and social skills to talk about the things which troubled me but a growing sense of concern for people and the environment merged into a vocation. The stories of people like Albert Schweitzer and Richard St Barbe Baker, who left comfortable lifestyles to help others, showed me a way forward.

Not knowing what to do, I prayed a child's prayer. "Father God, please use me somehow, somewhere, to make a difference."

I found deep satisfaction in growing things. I spent many hours in our large veggie patch watching, learning and helping my father.

CHAPTER FOUR
The call of the Sahel

My aversion to cities made the universities in Melbourne unappealing. I enrolled to study for a four-year Bachelor of Rural Science degree at the University of New England, in Armidale, New South Wales. The relatively small campus on the beautiful, rural New England tableland was a big drawcard. Professor Bill McClymont and his colleagues pioneered research and teaching in agricultural systems. Rather than siloing the different components of complex food and fibre production systems, the Rural Science course explored the interconnectedness of technical, economic, social and environmental parameters. This approach appealed to me.

I am by nature impatient and active, and not inclined to study. Throughout my student years I had a burning ambition to finish my education and just get out there to make a difference! Often, I had to suppress this yearning and force myself to concentrate. To this day, even reading about land degradation and the unnecessary human suffering it unleashes will stir a deep restlessness.

I found it difficult to settle into university life. The discipline of study was challenging and I was socially awkward. I was 700 kilometres from Myrtleford, lonely and homesick. Early in my time at the university, a member of the student Evangelical Union gave me a Bible. I got into the habit of reading it every morning and found myself regularly coming back to the words of Jesus in the Gospel of Matthew: "Ask and it will be given to you; seek and you will find; knock and the door will be opened to you."[8]

I asked myself, if this verse is true, why am I struggling? Why am I so miserable? Finally, in frustration, I cried out to God, "If your word is true, if you are all-powerful and if you answer prayer, then help me and I will go wherever you call me. But if you are not what your word says you are, and if you do not answer prayer,

Nothing but cotton! During a university excursion, Liz and I visited a broadacre monoculture farm at Narrabri. The epitome of modern agriculture.

then I am just being a hypocrite calling myself a Christian, and I will stop pretending." There was no audible voice from heaven, no lightning bolt, no signs or wonders. Instead, a deep sense of peace and assurance came over me. I sensed that I was not alone, that God did answer prayer and that he was already helping me. I certainly didn't become an A-grade student, but much of the stress and worry of academic life melted away. I started to make friends and enjoy country square dancing and the local roller-skating rink.

Occasional student antics were welcome tension breakers. Late one night as I was getting ready for bed there was a knock on my dormitory door. I asked who was there but received no answer. I opened the door slowly. I couldn't see a thing in the dimly lit corridor. Suddenly, hairy hands grabbed my left arm. I yelled in fright. Even though I immediately recognised the hysterical laughter of my friend, Fred Ghiradello, the adrenalin had already kicked in and I couldn't stop pummelling him with all the force of my free fist. Fred had somehow come into possession of a life-sized gorilla suit and thought it would be funny to say hello!

The university chapel provided a calm retreat from a packed and demanding university course, and a solid alternative to the poor lifestyle choices into which many students fell. The Bible teaching and Christian fellowship became very important to me. Visiting speakers at the chapel included missionaries and representatives of mission agencies. The older members of the chapel congregation, most of whom were university staff members living off campus, were generous with their time and welcomed students into their homes, providing friendship, home cooking and an escape from an otherwise artificial existence in the dormitories.

It was at this time that I made a break with my Catholic upbringing. Christian denominations were never important to me. I was intrigued that divisions are perpetuated even after death—you will find separate Catholic and Protestant sections in graveyards! I wasn't pitching one church tradition against another. For me the question wasn't one of "correct" church affiliation. Jesus' warning in the Gospel of Matthew is clear: "Not everyone who says to me, 'Lord, Lord,' will enter the kingdom of heaven, but only the one who does the will of my Father who is in heaven. Many will say to me on that day, 'Lord, Lord, did we not prophesy in your name and in your name drive out demons and in your name perform many miracles?' Then I will tell them plainly, 'I never knew you. Away from me, you evildoers!'"⁹ For me, the question was, did I know him?

In January 1977 I bought a copy of *In Famine He Shall Redeem Thee: Famine Relief and Rehabilitation in Ethiopia* by Enid and Malcolm Forsberg. I was painfully aware of the 1973–1975 drought and famine across the Sahel—the semi-arid realm spanning the African continent between the Sahara to the north and the savanna to the south—and I read the book from cover to cover.

> When the rains failed, the grazing lands withered and shrank. The nomads moved farther south. Those lands withered, too. Water, the most precious commodity of the desert and its borderlands, disappeared. The shallow wells of oases, and the mud holes of the rainy season water courses dried up entirely. Cattle died. People died, too. ... The damage that has been done to the Sahel is enormous. Some observers feel that much of it can never be repaired. The Sahara has been steadily expanding for decades, devouring more and more of the pasture and arable land on its rim. It is reported that in 1973–1974, the Sahara advanced as much as 60 miles in some places. ... There is no easy solution to the problems of drought and famine. To correct the misuse of the land and restore a proper balance of nature will be painful, slow and costly. Man must pay the price of abusing God's creation.¹⁰

I underlined this and other sections of the text and wrote comments in the margins. Mindless development at the expense of the environment was fatal. How was desertification and this unfolding human tragedy going to be turned around? This was my calling. I felt the adrenaline roar through my veins and an overwhelming compulsion to get up and go.

I then became aware of the work of Bruce and Norene Bond. Like Enid and Malcolm Forsberg, Bruce and Norene were Sudan Interior Mission (SIM) missionaries and had worked in Ethiopia for 20 years. I was inspired by the gentle and humble way this couple had encouraged the church and helped poor farmers grow more food. By reading about their experiences, I came to appreciate that development is not merely a matter of throwing money at problems

or having clever technical fixes. There are spiritual dimensions to poverty which money and technology cannot fix.

There are plenty of fish in the sea

Even though it did not appeal to me, I was prepared to remain single. I never expected to find an Australian girl mad enough to want to live with me in a mud hut in the middle of Africa. However, I couldn't help being attracted to one of the girls in my year, Liz Fearon. She sang like an angel in the chapel choir and was also a member of the student Evangelical Union.

It was vital to me that my future partner have her own sense of vocation. I had heard stories of couples who had moved overseas only to return home prematurely because they did not share the same calling. When, not if, problems arise in a foreign country, without a strong sense of purpose it is easy to become discouraged and overwhelmed by a desire to return home.

Liz grew up in Melbourne and attended Strathcona Baptist Girls Grammar School. Through her understanding of science, she had come to the conclusion that God didn't exist. One morning, the daily Bible reading for the school assembly came from 1 Kings 19. In the midst of a tempest, God speaks to Elijah in a still, small voice. From that moment, Liz knew that God was speaking to her in a still, small voice calling her to serve in some way. As her final year of school drew to a close, she considered following in the footsteps of her father and grandfather and studying medicine. However, upon opening a brochure on the Rural Science degree offered by the University of New England, she sensed that voice immediately, steering her to study agriculture with a view to serving overseas. At university she joined the Evangelical Union. As the missions representative she received information from a variety of mission agencies and arranged talks by visiting missionaries. Liz also was attracted to the holistic approach to mission (evangelism and meeting physical needs) she saw in the Sudan Interior Mission (SIM) literature.

The 1973–1975 famine in Ethiopia was not an act of God. It was the consequence of conflict and our violation of the environment.

We found ourselves in the same cohort, doing the same course. I was very shy, and it took me until the third year to summon up the courage to talk to Liz. To reach the

Early in our marriage, our faith was tested in small things. This helped us to trust God in the big challenges to come.

university campus from the residential colleges, most students walked along the affectionately named "goat track." After much mental rehearsing, one day I engineered it so that "by coincidence" Liz and I would walk back along the goat track alone. I asked her why she was studying agriculture. At that time the students were overwhelmingly male, so my question had a veneer of innocent curiosity. Without hesitation, Liz responded that she believed God was calling her to serve in agriculture in Africa! That was all I needed to hear. I couldn't have been happier. At this moment I knew in my heart that she would become my wife.

Being of practical Sicilian stock, Nanna had always told me, "An-do-ny, there's plenty-a fish in-da sea. Catch-a good-a-one! Make-a sure she can milk-a-da cow!" I had already done my homework. It had come out in conversation that Liz had milked a cow—only once, mind you, but that was good enough for me! Common sense told me that if a wife were to survive in Africa, she would need to be tough. One day, while we were hunting for botanical specimens near Lake Zot on the university campus, Liz lifted a rock and grabbed a writhing, hissing blue-tongue lizard by the neck. I was astonished. She easily passed the "not-squeamish" test as well.

Liz felt that she did not want to go to Africa as a single person. However, after wrestling with this challenge, she let go of her desire for a partner and told God that she was willing to go by herself. I didn't know about her struggles, but it was at this very time that I felt bold enough to try to get to know her better. My enthusiasm was not requited. Obviously, my skills in this business of courting required some honing.

In time, love blossomed. The turning point came with a rare stroke of brilliance on my part. Years before I had written a poem for a girl who subsequently showed no interest in me at all.

I retrieved the poem from its hiding place and gave it to Liz. There was something in that poem that revealed a depth that she had not recognised before. There is little of distinction that I can add to the great body of literature on falling in love. Suffice to say that in finding my soulmate I felt completed. Being loved and loving someone else brought an inner peace and growing confidence into my life.

SIM missionaries Anne and Paul Weekly came to speak at the university chapel. Paul had agricultural training and they were assigned to the Maradi Farm School in Niger. However, they eventually set off with a small baby to work with Fulani herders in Burkina Faso. Paul and Anne were open and candid about the joys and struggles of missionary life. Being taken into their confidence meant a lot to Liz and me, and the idea of going to Africa became more real for us. Liz and I signed up to receive their regular newsletter. We corresponded with them for many years and eventually met again when our older children ended up at the same boarding school as their children.

On the eve of our engagement party at Liz's parents' home, one of life's imponderable serendipities occurred. I was flipping through a family photo album before the guests arrived, when I came across a yellowing newspaper clipping. The article had a picture of 10-year-old Liz with her father, Dr David Fearon, holding a quail chick. Liz had brought him the chick which had appeared dead. Dr Fearon calmly laid it on the palm of one hand and gently stroked its breast until it revived. Thinking the story noteworthy, Liz's aunt sent it to *The Sun* newspaper. I recognised the article immediately. I had read it as a child. I liked the story so much that I cut it out of the paper and added it to my collection of curiosities. On seeing that article again, I remembered my reaction the first time I saw it: How great it would be to grow up in a family like that! Liz and I were married on 31 March 1979 and attended our graduation during our honeymoon.

Nanna was right about the importance of catching a "good one." Liz is my best friend and the love of my life. We share a common vision and work as a team to fulfil it. She has been a source of strength, encouragement and perseverance in the face of great challenges. She has stood by me when others criticised and undermined me. She deserves to fully share in every ounce of credit for what has been achieved. Without Liz, there would be no story to tell.

When we returned from our honeymoon, Liz and I worked for my father at Buffalo Farm Equipment and lived in a vacant share farmer's house on a tobacco farm two kilometres from Myrtleford. In the middle of our first night back from our honeymoon we were woken by the house shaking. The owner of the property had mischievously tied his horse to a nearby tree and it was scratching itself against the bedroom wall! Liz took charge. "You're the man! Go and see what it is!"

We spent a happy nine months in Myrtleford but, as time went on, we became more and more discouraged as we seemed further away than ever from going to Africa. We had hoped to obtain relevant work experience in Australia before going overseas but, except in the agricultural chemical and poultry food industries, there was little work available for new agricultural graduates. One day, at a particularly low point, Liz said, "Let's pray about it." We did. That very week, in the Christian magazine *On Being*, we saw an advertisement for the first ever Agriculture and Missions Conference. The event was to be held by the Agricultural Christian Fellowship and World Vision in Gilbulla, near Sydney. The magazine also contained a list of "opportunities vacant" with various missions. A number looked interesting, but we were unsure how to make a decision.

Gilbulla conference

At Gilbulla we were surrounded by people who were either preparing to go overseas or already had decades of experience. John Steward and other development practitioners and deep thinkers challenged us and helped us better understand our faith walk and the dynamics of international development work. These encounters had a pivotal impact on our approach to development. We were introduced to *Appropriate Technology* magazine, *World Neighbours* journals and the writing of Peter Batchelor (*People in Rural Development*, 1993), Juan Flavier (*Doctor to the Barrios*, 1970) and Roland Bunch.[11]

In his book *Two Ears of Corn: A Guide to People-Centred Agricultural Improvement* (1982), Roland Bunch shares the results of his research into the successes and failures of several decades of development projects. He demonstrated the importance of farmers taking responsibility for their own problems and finding the motivation to solve them. He emphasised minimising destructive dependency on foreign aid and encouraging farmers to experiment. It is tempting for expatriate experts to think that they have the solutions and that their job is to impart knowledge to those they are helping. At best, this fix can only ever be temporary. Agricultural systems are not static. Climate, growing conditions, markets and the arrival of new pests and diseases all change over time. Hence, solutions which are effective during the life of the project may be ineffective if conditions change, leaving farmers in the same state or worse off than before.

Roland Bunch believes that genuine development only occurs when farmers are enabled and empowered to help themselves.

As Roland put it, "If what you do does not last, you have done nothing at all." *Two Ears of Corn* became my development Bible. My only regret is that I didn't follow his advice more closely. We made mistakes, but I often wonder how much worse things might have been without this solid grounding.

Sudan Interior Mission, Sydney

From Gilbulla, we took the opportunity to visit the Sydney office of SIM (Sudan Interior Mission, later renamed Serving in Mission). SIM is an evangelical, interdenominational mission organisation. It was established in 1893 and is currently active in over 70 countries, working in church planting, theological training, medicine, agriculture, relief and education. I liked their holistic approach to ministry—acknowledging that a person is neither a soul-less body that only needs food, nor a body-less soul needing only salvation. We spoke with SIM co-directors Rob and Pat Brennan. This conversation led us to take a one-year intensive Bible in Missions course at the Bible College of New Zealand (BCNZ) in Henderson, Auckland. Why only one year? My impatience to get out there and do something again came to the fore.

The only time I ever saw my father cry was the day we left for New Zealand. Until that point, he held out hope that I would come to my senses and change my mind about going to Africa. When he came to Australia as a boy, he had to get to know his father, learn English and adjust to very different surroundings. Having first-hand experience of poverty, Nanna had drummed into him the need to be strong and become "somebody." Dad worked very hard to become successful and respected and it seemed to him that I would be relying on handouts. He was disappointed at not having me in the family business, and he was afraid of what might await us in Africa. Even so, he accepted our decision and went on to strongly support us. Mum took the news calmly. Perhaps in her thinking becoming a missionary was a close second to her fervent wish that I become a priest! Nanna had turned her back on poverty to create a better life for her family and felt that I was throwing away my education and wasting my life. In her very Sicilian way, she was upset with me and never came to terms with my living in Africa. Liz's mother, Jill, always thought her eldest daughter would do something unusual with her life. After the initial shock, she and Liz's father, David, supported us wholeheartedly. Fortunately, we were able to stick to our convictions, and hopefully the lives of many others are the richer for it.

Bible College of New Zealand

When we arrived at BCNZ in 1980, accommodation on campus for couples without children was limited. We were given a corner room which had been used for storage. Every day we swept, vacuumed and hung our blankets on the clothesline to get rid of the accumulated dust, which gave me red eyes, a runny nose and made me sneeze. My angry and grumpy reaction shocked me. Where was the valiant and dauntless missionary? I wondered about my suitability for life in Africa where I would experience far worse discomforts and injustices. I vowed to change my complaining ways.

Time and again, doubts would fill my mind. Who did I think I was? I lacked self-confidence and did not know if I could make a difference. But I took courage from Scripture: "For we are God's handiwork, created in Christ Jesus to do good works, which God prepared in advance for us to do."[12] This verse gave me assurance. I didn't know the future, but God did. Moreover, his word said that before I was even born—in fact, before the earth was made—he had me (and every human being) in mind; he had prepared good things for me to do. That was all I needed to know. Experience has taught me that just enough is revealed to us to enable us to take the next step. As J. P. Morgan is reputed to have said, "Go as far as you can see; when you get there, you will always be able to see further." That first step is a leap of faith. But from that new vantage point, one can see further and then be able to take the next step.

Formal classes at Bible college included subjects such as Old and New Testament, Anthropology, Sociology, and History of Missions. We often discussed the mistakes of past and present missions and the need to work in ways that were sustainable and respected local cultures.

Encounters outside the classroom were also significant for us. Many of the students, lecturers and visiting speakers had been missionaries. Many had faced extremely challenging situations and yet had remained faithful to their calling. This made a deep impression on us. It strengthened our faith to hear how they had taken leaps of faith in leaving their own country for foreign lands, how they had struggled, failed and triumphed in different circumstances, and how God had answered their prayers. Their stories of personal transformation kept our vision alive and encouraged us to persevere.

On Sundays, we attended Te Atatu Bible Chapel, where the Maori congregation warmly welcomed us into their close-knit community. This rich cross-cultural experience provided valuable lessons and insights that helped prepare us for the field. I threw myself into the church's social activities and youth outreach, and on special occasions I helped prepare the *hangi*—meat and vegetables cooked in a pit with pre-heated stones.

This was a time of growth and learning to trust God for all our needs, including finances. When we arrived in New Zealand, we barely had enough money to pay for our accommodation, food, studies and return ticket. Time and again, when we were down to our last dollar, someone would call the college to ask a student to do odd jobs. One time, Liz's mum was coming to visit, and we had no money. Two days before she arrived, we received a letter in the mail from our home church treasurer with a cheque and a note: "Apologies that I am only sending this now;

While studying at the Bible College of New Zealand, Liz and I were embraced by members of the Te Atatu Bible Chapel.

it has been ready to go for six months." The timing was perfect! Had the letter come any earlier we would have spent the funds and not had the means to host Liz's mum! These seemingly small things were having a significant impact on our weak faith. It was exciting to have our faith tested in small things. It was wonderful preparation for the times ahead, when we would have to continually trust God for big things.

While in New Zealand, we met my heroes Bruce and Noreen Bond, veteran missionaries who had spent 20 years with SIM in Ethiopia. It was through reading their stories while I was at university that I first became attracted to SIM. Initially, my admiration made me a little guarded with Bruce. He was a leader and authority figure, and I wanted to live up to the expectations I assumed he held. But Bruce and Norene were warm and welcoming and treated any prospective SIM missionary as family. They knew what it was like to be separated from family and caught up by the uncertainty and excitement of preparing to move overseas. Bruce was now the national director for SIM New Zealand and also on the BCNZ board. He asked after us when he attended meetings on campus. He and Norene had lunch at the Bible College each month to chat to students. They were great story tellers. Again and again we were enthralled hearing about their adventures in Ethiopia with their children, Ethiopian colleagues, other missionaries and the communist government.

Stories about emboldened and empowered village folk were the most memorable. One afternoon Bruce and the village elders were summoned to appear the next morning before the government administrator in Dilla to answer charges made against them. The total population of the Gedeo people at that time was approximately 500,000. The believers numbered between six and seven thousand and were growing rapidly.

The administrator called us into his office and after the normal greetings began telling us about the unrest that was being caused by this new religion the foreigners had brought to Dilla.

"There is a lot of fighting among the people," he said. "This religion is dividing families and communities."

Because of a whole string of false accusations, he ordered us all to stop preaching and teaching and to return to the previous ways. I wondered how we should answer this man who was genuinely seeking to maintain peace in the region. The oldest man in our group, illiterate and almost blind, stood up.

"Sir, are you not aware of the changes the gospel has already brought to our community?" he asked. "Those of us who have believed in Jesus have given up our old drunken ways. Before we believed, when we sold our coffee crop, we would go into the drinking houses around the markets and drink until we were drunk. We would fight amongst ourselves and go to the prostitutes. We would go home and beat up our wives and children. Because our money was spent in drinking and at the brothels, we would live in poverty and squalor for the rest of the year. Our kids never went to school, never had any decent clothes, never had much medical help. And our houses were leaking and rotting."

"Now, since accepting this so-called 'new religion,' our way of life has completely changed. Instead of wasting our money as before, we are now able to build good houses with tin roofs. Our children are becoming literate. We're able to go to the clinic for medical help. We are able to dress ourselves in decent clothing, as you can see today. Furthermore, for the first time we now know that God loves us and sent his son, Jesus, to be our saviour. Our sins have been forgiven and we are children of the heavenly king. There is no way we will ever give up preaching and teaching this new religion, no matter what you do or say to us."

> Silence. God had indeed used the weak of this world to confound the wise. The governor knew that what the old man said was true. Fearing further rebuke, he let us go with a word of caution.[13]

Bruce and Noreen invited us to their home for meals and we were able to share our fears and struggles. We told them how difficult it was to study while not formally accepted into the SIM family. This left us feeling insecure about our future. Bruce contacted the Australian SIM director and intervened on our behalf. The outcome was that we were accepted as SIM missionaries by the NZ SIM council while we were still at Bible college.

Destination Niger

Then the decision had to be made as to where we would be assigned. I had written in our SIM application form that I was interested in dryland agriculture. Bruce Bond favoured Ethiopia but the attitude of the communist regime made that unlikely. Our Victorian SIM representative had been to Ghana and was keen for us to serve there. The Australian national office persuaded us to consider a post at the Maradi Farm School in Niger. However, when I realised that we would not only have to learn to speak Hausa, the local language, but also French, the national language, I pleaded the case for Nigeria, Niger's English-speaking neighbour to the south. Finally, we received a phone call from the international head of human resources who personally requested that we reconsider Niger. Pat and Evelyn Franje, the missionary couple managing the Bible school and rural development project in Niger would be retiring the following year and there was no one to replace them. He cited my interest in dryland agriculture. "You can't get much drier than Niger!" We wanted to go where we were most needed, so we agreed on Niger.

From Auckland we flew directly to Sydney to complete a Summer Institute of Linguistics 12-week summer intensive at the University of New South Wales. There, we were immersed in anthropology, grammar, phonetics and language-learning techniques. The most useful part of the course for me was the introduction to an approach called Language Acquisition Made Practical (LAMP). The schedule was brutal. Liz found the challenge and the analysis of language

While in New Zealand, we met my heroes Bruce and Noreen Bond, veteran missionaries who had spent 20 years with SIM in Ethiopia. I was inspired by the gentle and humble way this couple had encouraged the church and helped poor farmers grow more food.

exhilarating, but I found it overwhelming and wondered if I would ever be able to string two sentences together in a foreign language. The anthropology and how-to-learn-a-language subjects laid a strong foundation for us in terms of our attitudes, techniques and approach to learning both language and culture.

Five weeks before the birth of our first child, we returned to Melbourne and stayed with Liz's family. We still needed to earn a living. A family friend and assistant director of nursing at the Austin Hospital in Melbourne alerted me to their need for an orderly in the spinal-injuries unit. What an eye-opening experience! In a single moment lives could be changed forever, be it through an act of foolishness or pure accident. The drunk driver involved in a car crash and the innocent housewife struck by a brick falling from her own carport roof were both equally sentenced to a life of restricted mobility. All ages were represented—children and young adults with their whole life before them through to the elderly who had been looking forward to enjoying their retirement. Remarkably, after an initial period of denial, grief, anger and depression, many patients went on to live meaningful and fulfilling lives. However, others never accepted their situation. They never seemed to move on and fell deeper into depression and bitterness. My job was not complex. I was in a team of three orderlies who had to turn patients every two hours to prevent the formation of bed sores. I took a personal interest in each patient and tried to connect with them as individuals.

While in Melbourne we were also buying up supplies which might not be readily available in Niger, such as clothes, kitchen utensils and children's toys. We filled four tea chests with items to send unaccompanied. We also needed to raise a team who would support us financially and through prayer. During the nine months leading up to our departure, sometimes alone, sometimes with Liz, I regularly spoke in churches and to small groups in country Victoria and New South Wales, Melbourne and Sydney. Because SIM does not have an independent source of funds, missionaries depend on pledges made by hundreds of individuals and churches. Their faithful and sometimes sacrificial giving and encouragement made it possible for us to live and work in Niger for 17 years.

I dreaded public speaking. I lacked self-confidence and experience, and I found this aspect of mission very difficult. I was afraid of making a mistake or not having the right answer to a question. One morning I woke up with the words "Don't be scared; be prepared!" Later, a friend gently told me that if I am genuinely telling my own story, it can never be wrong. It is simply my story. If I'm asked a question that I don't know the answer to, I can simply say, "I don't know." There was no need to fear. The wisdom of these words made perfect sense. Rising to the challenge of speaking in public helped to hone my skills and, in the end, we were able to raise the necessary funds. This money went into a shared funding pool; if in any one month one member of the team did not have 100% support, they would still be guaranteed an allowance.

Today I speak regularly in large and small public forums around the world, and I often find myself speaking to journalists or into a microphone or in front of a camera. Before you can run, you must learn to walk. Before you walk, you learn to crawl. What had been extremely challenging for me has become routine. I have grown in confidence by listening to good speakers and through practice.

On 14 March 1981, I worked a double shift at the hospital and returned home at midnight. I had a snack and fell asleep exhausted. Half an hour later Liz woke me, "Get up! It's time!" Benjamin John Rinaudo was born at 8:00 a.m. the next morning and our lives were changed forever. We had no idea how much a baby would inspire love in our hearts or how much work raising children would involve. A couple of months later, we returned to Sydney and did three months intensive French study with the Alliance Française.

On 26 September 1981, our families, some close friends and mission representatives saw us off at the airport. It was very difficult for our families. After an intense year, I was relieved that we were finally leaving, yet sad to be saying goodbye. Our life experience had pointed us in this direction. It was strange and yet not strange, frightening and exciting. We were simply going as far as we could see, so that we could take the next step.

CHAPTER FIVE

A welcoming smile is better than a welcome mat

Liz and I had never travelled further than New Zealand. Were there ever any less likely globetrotters? In October 1981, we flew with Alitalia from Melbourne, via Rome, to Kano, Nigeria. The moment I buckled up my seat belt, the audacity—perhaps the stupidity—of what I was doing struck me. Had I gone mad, leaving the comfort and security of family and country to take my young wife and six-month-old son to Africa? Was my father right to be so worried? After a service in our home church, I overheard someone talking about us. "It's okay for Tony and Liz to go off and do their own thing, but they should think about their son!" That stung. But there was no turning back.

The plane from Rome to Kano was packed with Nigerians returning home with their bulging sacks of shopping. Having Ben with us meant that we also had a lot of gear, and so we decided to wait until everyone else had left the plane before we moved. When we stepped onto the tarmac we were immediately struck by the heat. How were we going to live in this climate? We were last in the slow procession. By the time we reached the immigration desk, the plane had reloaded and taken off. That was just as well. As the official thumbed through our passports, his demeanour darkened. "No Nigerian visa!" he exclaimed. I said that we had been told we would not need one as we were in transit and proceeding to Niger the next day. (We had no visa for Niger either! We had also been told that we could obtain a visa for Niger in Kano.) I was hot and exhausted and not prepared for what the immigration official said next. "You have broken the laws of the Federal Republic of Nigeria! You can get back on that plane and go back to wherever it is that you came from!"

Hausa people are curious, generous and hospitable. They feed guests first even if they have to go hungry themselves.

Normally, I would have panicked, but a deep sense of peace and assurance came over me. Besides, the plane had already left! At this point, our mission's business agent arrived and very calmly spoke to the official, assuring him that a SIM-Air plane would take this troublesome pair off his hands and fly them to Niger the next day, with or without visas for Niger.

The official listened to our business agent respectfully and glared at me but when his gaze fell on Ben he melted and his anger vanished. Smiling, he waved us through. That night in the church guest house I reflected on the day's drama. This was God's way of reassuring us. If God could get us out of a mess of my own making, then we could also rely on him for whatever lay ahead in Niger.

The next day the mission pilot took us to the Niger consulate. Normally it takes three days to obtain a visa and the following two days were public holidays to commemorate Nigerian independence. The pilot explained our predicament to the consul who agreed that, indeed, we were in a difficult situation. The visa was granted within three hours, and we were on our way that same day as promised. Seventeen years later we would be farewelled by the same man.

We were welcomed to Niger by our friends Kim and Bart Vanden Hengel. They lived in the town of Tsibiri in Maradi Region. Kim had been at university with us. As we drove into Tsibiri we immediately became aware that this was a place where everyone kept livestock. It was the end of the wet season and the ammonic smell emitted by the moist soil transported Liz to the shearing shed where she had worked as a rouseabout (unskilled farm labourer). We were told to keep our mosquito nets on our beds and not to get out of bed barefoot in the night. We had to shake our shoes before putting them on and wear them to protect our feet from scorpions.

We then flew to Galmi. To our surprise, at the SIM hospital Nikki Knowlton, the administrator's wife, gave us clothes for Ben and enthusiastically welcomed us with cold Coke and home-made ice cream—luxuries we thought we would not see again for four years.

Madaoua

From Galmi, we were driven to Madaoua—in 1981 an important town of 4,000 people—for four months' Hausa language study. At Madaoua, the principal east–west highway crosses the road going south to Nigeria and north to Tahoua, an administrative and commercial centre.

Our home for four months while at language school: rendered mud brick walls, concrete floors, bars on the windows and lots of mosquitoes.

Apart from newer government buildings and the homes of the wealthier inhabitants, most houses were made from mud brick with roofs of either iron or mud (which needed to be re-rendered every other year). The dusty streets were lined with leafy Indian neem trees, which had descended from two trees planted 29 years earlier by our hosts, Alvin and Lolita Harbottle.

In Madaoua, we lived on the mission compound in a cement-rendered, mud-brick house. It had a corrugated iron roof, power and running water. I asked Alvin why all the insulated electrical wiring was exposed on the inside walls. Surely it would be tidier if it ran up through the brickwork? Alvin said that he'd had an argument with the electrician, who was on short-term assignment, and Alvin had won: if there were a problem, the electrician wouldn't be around to fix it, so Alvin wanted to see and to have access to the wires. Fair enough. Of solid rural Canadian stock and with a very dry sense of humour, we never quite knew when Alvin was being serious and when he was pulling our leg. Our classroom was painted with three shades of green in not-quite horizontal bands. "How did that come about, Alvin?" Alvin, whose sole mission in life was to share the gospel, replied, "Actually, I'm an ace handyman but, out here, unless you want to be assigned full-time to maintenance, it's better to not show any level of competence!" What might have been closer to the truth is that, being thrifty, he didn't buy enough paint at the outset. He started painting from the floor up on all four walls simultaneously and ran out of paint twice. With each new purchase, he couldn't match the colour. Nevertheless, let's stick with Alvin's version of events.

Alvin met Lolita, our Hausa language instructor, at Bible college in Canada. He was convinced she was the one, but she would have nothing of it. Her calling was to serve the Lord

Liz and Ben in the kitchen in Madaoua. Until we were able to haggle in the local market, we bought food in cans at set prices.

in Africa, not to look after a husband and raise children! Lolita graduated before Alvin and sailed to Nigeria alone. Alvin, if nothing else, was persistent. At each port of call, Lolita was greeted by a letter containing a fresh marriage proposal. Finally worn down, she relented on one condition: they would take turns looking after children and sharing the gospel in the villages. And that's how they operated. Alvin would go trekking into the surrounding villages for two weeks while Lolita looked after their children and, when Alvin came back, Lolita would go out. They lived in Niger for 40 years. It was 35 years before they saw their first convert.

It was crucial that we learned the vernacular before we took over from Pat and Evelyn Franje, who were managing the Maradi Farm School and reforestation project, and supporting the Preparatory Bible School. We were to have a short overlap with the Franjes to learn the ropes, and then we would be more or less on our own. Very few locals spoke any English and we had little French, so we were under pressure to become proficient in Hausa as soon as possible.

Our time in Madaoua was frustrating. We rarely got a full night's sleep which made concentration in classes difficult. Nights were hot. Hungry mosquitoes managed to find even the smallest holes in our bed nets. Most nights, Ben woke many times. When we got back to sleep, either braying donkeys, barking dogs, or calls to prayer from the competing mosques would soon wake us again. We were exhausted.

We wanted to be out amongst the people and practising new phrases. But this was discouraged by Lolita, who was protective and didn't want anyone to take advantage of us. Our antiquated Hausa textbook did not use everyday phrases to illustrate the various aspects of language. This made it difficult to relate to the language or use what was being taught. Fortunately, the LAMP methods we had learnt in Australia made all the difference: diligent practice each day, creating lists of words and phrases, recording and playing back sentence drills, and carrying a pen and notebook everywhere we went to collect new words and expressions that we heard. LAMP also taught that it is normal to reach plateaus in language learning and this encouraged us to keep working even when progress seemed impossible. Through perseverance, we would eventually reach a new plane of understanding which would in turn open the doors to greater comprehension and expression.

It is important to be able to laugh at yourself. Even children laughed at our inability to comprehend or say the simplest things. Rather than feeling cross or retreating, laughing with them broke the ice and paved the way for ongoing exchange and learning. We knew we were making progress when feedback transitioned from "There is no Hausa" to "There is Hausa" and when we were complimented with, "You speak Hausa as well as a Kano donkey."

One feature of Hausa culture that I love is the frequent use of proverbs that hold wisdom, moral instruction and lessons about what is ironic, predictable or inevitable in human behaviour. Animal caricatures feature heavily in Hausa proverbs.

Jaki ya ga toka, sai birkita.
When the donkey sees the ash pile, he will automatically take a dust bath.

Hutun jaki, da kaya.
The donkey rests with its cargo on its back. (Even when a donkey is resting,
the owner leaves the cargo on its back. From some burdens there is no rest.)

Albarkacin kaza, kadangari ya sha ruwan kasko.
The blessing of the chicken is that lizards get to drink from their watering bowls.
(What benefits one can inadvertently benefit another.)

Likewise, the tentative, back-and-forward, rocking motion of chameleons as they walk provides a lesson for the brash and reckless: *Tafi a hankali cikin duniya, in ji hawainia.*—Travel cautiously in the world, says the chameleon.

The older generation peppered their conversation generously with proverbs. They would cite the first half of the proverb, "*Jaki ya ga toka*," and the listeners, with knowing nods, would automatically recite the response, "*sai birkita*." By just mentioning the first part of the proverb the meaning is understood and the interlocutors are spared lengthy explanations. Sadly, I found that many younger, educated Nigeriens did not know or understand the majority of proverbs.

African soundscapes

From my desk in Madaoua I could observe the comings and goings in the street which ran along one side of our compound. Undetected, I watched and listened to the daily routine, from the *gari ya waye* (awakening of the town) to the silence of night which was punctuated by the occasional dog fight or domestic drama.

Five times a day, the Muslim faithful were called to prayer by *muezzins* with loudspeakers in minarets. Soon after the early morning prayers, with barely enough daylight to see, the hired shepherd appeared, calling as he walked through the town. Rusty hinges creaked as the wood-framed, corrugated-iron doors of the high-walled compounds were flung open in step with the passing of the shepherd and swung shut once the animals had joined him. From each household an assortment of sheep, goats and cattle filed out taking their position in the growing herd and swelling the cacophony of hoof beats, guttural gurgles, bleats, baas and bellows. Any household which had not paid the modest shepherding fee saw their animal lightly bonked on the head and sent back. The shepherd kept mental tabs on the hundreds of animals in his care and which owners had paid up and which hadn't.

The shepherd would return at dusk. Ewes, nanny-goats and cows, sporting bursting udders, bleated and bellowed, calling out to their hungry offspring at home which responded with equal enthusiasm. Each animal knew its own home and automatically peeled off from the herd at a gallop to be let into the compound by the owner, to drink and be greeted by their eager young which butted their udders while pulling and sucking voraciously on the teats. What sustenance these sleek animals gleaned from the arid landscape was a mystery. When they returned home, the women gave them grass hay and the water they had used to wash the grain. The older the grain, the more weevils. The more weevils, the more fine-grained powder and weevil parts washed into the off-white slurry. The thicker the texture, the more nutrition was ingested by the animals drinking the wash water. *Kumzo*, the dry, porridge crust prised from the cooking pot, was also fed to the livestock.

Of all the animals kept in people's compounds, donkeys were the loudest. They certainly had the most to complain about. Much of Niger's economy rests on the back of this much-maligned beast of burden; yet, it is often malnourished, abused, beaten and overloaded beyond all reasonable limits. And, for all its hard work, the donkey is considered lazy and stubborn. No wonder they ratchet up their squeaky hacksaw braying for all the world to hear: Eehaw! Eehaw! EEHAW!

Once the livestock left in the morning, the clanging of metal buckets and the rising voices of women and girls could be heard as they waited in line to collect water from a communal tap. This is where women could catch up on gossip and air their grievances away from the scrutiny of their menfolk. Even very young girls were tasked with carrying water. Thinking myself strong and fit, I once foolishly tried to lift a traditional clay pot full of water to balance and carry on my head the way they did. Pots full of water weighed 15 to 20 kilograms. Let's just say that I never tried to do it again. My admiration for the strength, poise and grace of Nigerien women and girls knew no bounds.

With water, the women could begin meal preparation. Millet was the staple food for most Nigeriens. It was most commonly eaten as *tuwo* (a heavy porridge) or as *furra* (a thin gruel). In bigger villages and towns where there were small, diesel-powered mills, women queued up to have several days' worth of grain milled into flour. The Lister motors were started by turning a crank handle. They were stiff and difficult to turn at first but became faster and easier as the momentum of the flywheel supported the motion. The engine produced a characteristic sound similar to that of clapping two hardwood slats together, slow and distinct at first, then getting faster and faster until it was chugging on its own.

One of our language school assignments was to spend the morning with Madai, the pastor's wife who lived on the compound. Liz recounted in a letter home:

Madai took the grain to a machine to have the bran removed, brought it back and washed the dirt and remaining bran out, then she took it to another mill to grind it into flour—each time having to queue up. After returning home again, Madai sifted it, pounded it with pestle and mortar, and gradually added water until it formed grapefruit-sized balls of paste. As she cooked each ball, she repeatedly pounded it while adding buttermilk. Then Madai put it in a bowl and gradually added more milk until it was like a thin gruel. This process took from 8:30 a.m. to 12:30 p.m. and she does it every second day.

In the bush they would have to hand-grind the grain with pestle and mortar, and remove the bran manually. Any spilt grain or flour was quickly raided by scurrying chickens, goats and pigeons. The cooing of pigeons, both wild and domesticated, was ubiquitous.

Pounding grain, using a deep wooden mortar and pestle, produced a rhythmic thump-thump sound. Better still was the beat generated by three women working the same mortar. Each person pounds the grain in turn. The first person pounds the grain three times and on the third upswing, releases the pestle into the air while clapping twice in counter point to the thump, thump, thump. Meanwhile the second person positions herself and begins pounding as the first person withdraws her pestle. When she is finished the third person moves in, and so it continues until the grain is adequately pounded: thump, thump, thump, clap, clap, thump, thump, thump, clap, clap … Even in toil there is rhythm and music. With the flour pounded, the fires would be lit, and this was followed by the banging of aluminium pots being positioned on the typical three-stone hearth. Then would come the tinny, hollow sound of water pouring onto metal and the clanking sounds of large wooden spoons hitting the sides of aluminium pots as the meal was stirred. Aluminium pots—a modern innovation over clay pots—were locally fabricated by artisans who melted down old aeroplane and car parts and poured the molten metal into sand moulds. To my astonishment they often worked barefoot.

After they had washed, dressed and eaten, the men would begin to emerge and go to their workplace, the market or their farms.

Drawing water from an open well is strenuous and dangerous. It is a task performed daily by women.

At the start of the rainy season, they would rise earlier and trek out to the farms to sow with a small sack of seed in their hand and a long hoe on their shoulders. Children created their own distinctive din of high-pitched laughter as they met their friends on the way to school.

During the day, an armada of bush taxis, mini-vans, donkeys and carts delivered people and cargo to the crowded marketplace in the town centre. The buzz of people haggling and talking filtered into our language-school classroom. Goods traded at the market included hand-woven mats, rugs, home-made mattresses and pillows, furniture, grain, fresh fruit and vegetables, watches, radios, sunglasses and second-hand clothes.

A much bigger market was held on Saturdays at Tumfafi, about six kilometres out of Madaoua. It was the largest grain and livestock market in Niger. Practically every household in the town kept animals, so vast quantities of hay were purchased. Every Saturday, we would see long camel caravans passing our house and boys with heavily laden carts beating their donkeys to go faster. Only the heads and tails of some donkeys were visible under enormous loads of grass hay. To protect the privacy and security of local compounds, by law, camel riders are obliged to dismount from their elevated perch as they enter towns.

Children—usually girls—carried trays on their heads with snacks for sale. *Almarjirai* (Koranic students) chanted verses in Arabic to the faithful in the hope of being given food or money. Other children—usually boys—led the blind to the market to beg. Sometimes alone, sometimes in small groups, the blind, the lame, the deaf cried out for alms. "For the sake of Allah! For the sake of the Prophet!" When the boy alerted the blind person that we were white, the cry would immediately change to, "For the sake of Jesus!" Pungent spices were poured onto mats for display. As people passed by, sneezes would punctuate the general din of people haggling and talking, and of animals and birds bleating, mooing and cackling. The medicine men played a flute and sang out as they passed. Among their many wares was a "miracle" cure in a bottle that they claimed could fix arthritis, backache, toothache, headache and more. Liz and I were happy to stick with aspirin.

I salute women in rural Africa. They work from sunup to sundown, fetching water and fuel wood, cultivating crops, tending livestock, preparing meals and raising children.

After grain, the biggest import into the town by volume was firewood. From distant, receding forests, it came on 20-tonne trucks, donkey-drawn carts, the roof racks of bush taxis, the backs of donkeys and camels, and on people's heads.

At dusk, the action was played in reverse. The market gradually grew silent, except for where passers-by mingled near the food stalls of "sons of tables," merchants who had no shop but sold their wares from portable tables. Boys raced their empty carts back to Madaoua, beating their donkeys mercilessly.

We adapted quickly to the narrow range of ingredients available for preparing meals. We had been warned about this. However, we had not been warned about the time it would take to find the things we wanted at the market.

In our first weeks, with limited Hausa language, going to the market and bargaining for necessities was daunting. In fact, we didn't buy red meat at all during our four months at language school. The combination of the fly-covered meat, the difficulty of working out how to ask for the cuts we wanted and the challenge of cooking the often-tough meat meant that we ate a lot of canned mackerel—so much so that we can still hardly look at mackerel. The only fresh vegetables were pumpkin and local spinach. Before we came up with recipes we liked, we relied on canned peas and beans. In Maradi, there was greater diversity, as vegetables were brought over the border from Nigeria. Liz learnt how to make local sauces and soak vegetables in dilute chlorine solution so that we could eat them raw. We acquired several cookbooks written by expatriates with good recipes using local ingredients.

Maradi Farm School

Inspecting a well-protected village nursery with (right to left) John Ockers, Pastor Cherif and a village extension agent.

Following four months in Madaoua, we moved to the Farm School in Maza Tsaye (meaning Men Standing) on the outskirts of Maradi. Maradi is the administrative centre of the Maradi Region and the second largest city in Niger. Here we met Pat and Evelyn Franje whose responsibilities we were to assume.

My duties would include overseeing a preparatory Bible school and managing the property, which included an orchard, nursery and dairy. I was also to supervise the Maradi Windbreak and Woodlot Project. This was a reforestation program started by SIM missionaries following severe famine in Niger in the early 1970s and funded by the Canadian government. The SIM project manager could see that ongoing deforestation would make drought and famine more likely in the future and asked the donor, the Canadian International Development Agency (CIDA), for permission to use unspent funds to establish tree nurseries. CIDA agreed and later granted ongoing funding for the project I inherited in 1982. Project activities included raising seedlings grown either in a central nursery or in village nurseries and working with communities to plant, nurture and protect the trees.

When I arrived, there were three full-time project staff, three teachers, a farm assistant and a guard. At its height, the project employed 10 salaried local staff and varying numbers of part-time extension agents, who lived in the villages and were expected to practise and demonstrate on their own land what they taught others. Over the years many long- and short-term SIM missionaries supported the project in administration, community health and agriculture.

Maradi Farm School manager

As somebody who avoids conflict, finds it hard to say no and is reluctant to lead, I was ill-equipped for my role as manager. I was often anxious, fearful, angry and grumpy and this was affecting my relationship with Liz, who felt cut off as I retreated into my shell. The weight I felt with this responsibility was not eased by a very difficult staff member who, apart from being rude and demanding, turned out to be dishonest and a sexual predator. On 8 April 1982 I wrote:

> *Babba juji ne, kowa ya zo da shara sai ya zubar*. (The chief is like a dust-heap, everyone brings their sweepings and dumps them!) I am finding being in charge and making decisions hard. Being interrupted all the time and keeping track of money handed to me at all times of day for different things is hard. People want me to help them and try to get the most out of me or take advantage of me. People want loans but I don't know what limits or terms to apply or how they will repay. Isaiah 26:3, "You keep him in perfect peace

whose mind is stayed on you, because he trusts in you." Lord, please help me to keep a level head, keep calm and be mindful of you. I commit my way to you and trust in you.

Was there ever less suitable material for the making of a missionary? Each time I found myself questioning my calling, I recalled all the events and ways in which God had led me, and I concluded that I was in the right place. I needed to trust God and persevere.

Sarkin Hatsi

Pat Franje took me through the virtually treeless farmland to visit the dozen villages participating in the project at the time. The most distant village was 30 kilometres away.

Each village name told a story. Village wells, which could be as deep as 100 metres, were traditionally dug by hand and not lined. In time, the earth wall at water level would collapse, reducing the available reserve of water. At great personal risk, someone would have to go back into the well to clean out the silt. There was no equipment or crew on hand to rescue them. Unchecked, a sizeable and dangerous cavern could form. The fading cries of women and animals swallowed by the collapse of one well gave the name—Sha ka Fada (Drink and fall in) to one unfortunate village.

Some villages had more lighthearted names like Hannu ga Zanni (Hand on her skirt). My favourite name was Batafadua (Unmentionable). Many villages were named after long-gone trees or forests: Kuka Biyat (The five baobabs), Kalgo (Bauhinia), Tsamiya (Tamarind), Gao Goma (The ten gao trees) and Kurmi (Forest).

At dusk, after inspecting the fields, Pat and I sat on handwoven mats around a small fire chatting with the chief and a few villagers. By this stage I could speak Hausa haltingly and could catch the gist of some of the conversations. Like farmers everywhere, they spoke of the weather, last year's crops, their hopes for the coming season, of people who were sick or travelling or not heard of for some time. They teased each other, told tall stories and enjoyed each other's company. Time stood still. I felt happy. I felt at home. This was the good work that God had prepared in advance for me to do!

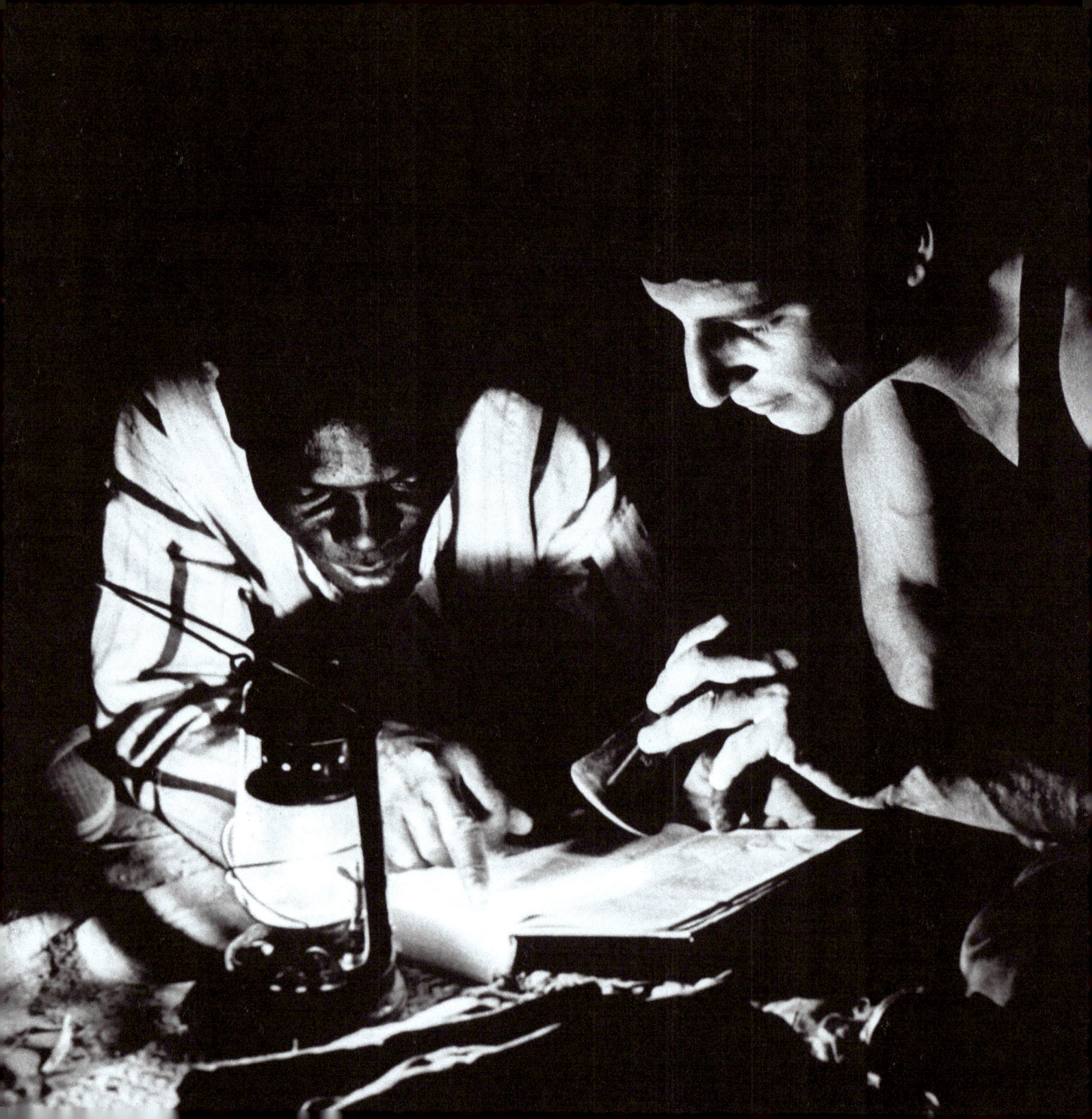

Asking questions was considered nosy and foolish. But if I didn't ask questions, I would never know or understand the people I came to serve. I was an outsider and knew nothing of the culture and practices. I was like a child, and children learn by observing and asking lots of questions. Fortunately, people were generally happy to answer my many questions. In hindsight, I realise that not knowing all the answers and asking lots of questions, rather than barging in as the grand expert, was actually very important for my acceptance and for building trust. When you ask somebody a question you are acknowledging their wisdom and authority. They become your teacher.

Hashimou, the young chief of a village called Sarkin Hatsi (Chief of millet), encouraged my curiosity. One day as I walked with Hashimou in his village, I heard muffled voices coming from the ground beneath me! I was intrigued. "The earth is speaking!" Hashimou laughed and took me to the entrance to a dugout which was a metre deep and just wide enough for the two men inside to sit side by side. The ceiling consisted of sticks covered with soil, which in turn had a tangle of thorn branches placed on top to keep animals—and curious visitors—from falling through. The men were weaving rope from the leaves of the doum palm (*Hyphaene thebaica*). "Why are you weaving underground?" I asked. "Well, in the dry atmosphere, doum palm leaves become hard and brittle. Apart from cutting our hands, the dry leaves break easily and are difficult to weave into rope. We keep the leaves moist in a bag. In this dugout, our breath keeps the air humid enough to stop the leaves from drying as we weave."

Walking a little further into the village I gagged at a repugnant odour coming from a family compound. It smelt like a dead animal which had been in the sun for days. "What's that!?" The housewife had made a black paste from ground and fermented roselle seeds (*Hibiscus sabdariffa*). She was forming the paste into patties and placed them on the roof to dry. I was shocked to learn that she was preserving *daudawa*. *Daudawa* is a valued ingredient which adds flavour and protein to normally meat-less meals of millet porridge. It is used in moderation and, after cooking, tastes—and smells—good! Well, if an Australian can

*Mai tambaya ba shi bata,
sai dai asirinsa ke tonuwa.*
(He who asks doesn't get lost but his ignorance is revealed.)
Hashimou and I read the Bible together in Hausa.

eat and enjoy Vegemite, a black, sticky, fermented by-product of the brewing industry, why not *daudawa*?

That evening I sat with a handful of village men on handwoven mats. There was no fire as would be customary—wood was too scarce. A single kerosene lantern placed in the centre of our huddle cast giant shadows behind us. Household livestock had been fed and watered and were settling down for the night. In the darkness, sounds became prominent. Children giggling and chasing each other, crying babies, pots and pans clanging, and the hum of adult conversation.

Resting in Sarkin Hatsi after a long day working with farmers in their fields.

Few words were exchanged while we ate the heavy millet porridge and vegetable gravy from a common pot. But as soon as the meal was finished the village din was drowned out by tall tales, gossip, banter, ribbing and raucous laughter.

"Tony, are you related to Tony Blair?"

"Tony Blair? Ha, ha. No, I am not from England."

"Where do you come from?

"Australia."

"Oh, Austria! Are the udders of your cows really as big as the ones on the powdered milk tins?"

"Not Austria, Australia! And yes, dairy cows have been bred to produce lots of milk and they really are that big."

"Can you bring some of those cows to Niger?"

"I don't think they would survive in Niger."

"Where is Australia? Is it further away than Mecca?"

"Yes. It takes six hours to fly to Mecca, but more than 24 hours to fly to Australia."

"Mamaki! (Amazing!) Did you hear that?"

I looked up at the African night sky in all its sparkling glory. "Australia is at the far end of the world. If you go any further you will fall off!"

"Did you hear that? If you go any further, you fall off the edge of the world! Listen to Tony! He speaks Hausa like a Kano donkey!"

"Do people in Australia speak Hausa?"

"No."

"Mamaki! How can that be? Do you mean to say that there is a country where people don't know Hausa?"

"Yes."

"But how do they get on?"

"They make do as best they can."

Heads shake in pity and amusement. "How many wives are you permitted to have?"

"One wife."

"Sannu! (So sorry!) Just one? How do you manage?"

"It is trial enough to keep one woman happy. How would I manage with two or more?"

Raucous laughter.

"It's not worth the trouble!"

"When our village chief turned 16, his father, the old chief, presented him with two wives!"

I turned to Hashimou. "Is that true? What was that like?"

Hashimou shook his head. "I was so young, I didn't know which way to look."

This was, evidently, a long-standing source of amusement and the men again erupted in laughter. "Well, Tony, if you ever change your mind …"

"Liz would lock the door! I wouldn't be able to return home!"

"Then you can stay with us!"

Moussa

Moussa approached me for work in mid-1982. He had been retrenched from his job and could not pay his rent or feed his family. I turned down most such requests; there were just so many

Extension agent, Moussa (left), shows a government forestry officer a village nursery. The barren landscape betrays the futility of our task.

of them and I had a limited budget. However, there was something about Moussa which appealed to me, and I sensed that he was genuine and willing to learn and to work hard. Importantly, to me at least, he was not arrogant. His lack of tertiary qualifications or expertise in the field of agricultural or rural development would have automatically disqualified him for employment with most Non-Government Organisations (NGOs). I was looking for character and teachability rather than academic qualifications. Many educated people looked down on uneducated villagers, considered themselves above living in a village or doing manual work and were unteachable. Not exactly a good recipe for influencing anybody.

I initially placed Moussa in the village of Waye Kaye (Enlightened ones). For their part, the village pitched in to build him a mud-brick house and provide land for him to farm. Moussa proved to be an energetic, enthusiastic and a charismatic leader. He had the ability to engage people and inspire them to try new things.

I expected all my staff to practise what they preached. Moussa didn't disappoint me. When we were promoting fuel efficient stoves made from mud, his home was the first in the village to boast one. His compound was encircled with a band of vigorously growing acacia trees which he planted and watered—a welcome respite from the other mostly scorching, shadeless compounds in the village. He even implemented my wildest ideas. Open defecation was the norm and constituted an eyesore and grave health risk. I asked Moussa to construct a bush toilet on his farm so at least the deposits would be buried and less likely to be discovered by flies and human feet. Of course, people used the facility—it provided a modicum of privacy after all—and the nearby millet plants were the best in the district. However, Moussa became the butt of so many jokes ("Moussa prays for us to get diarrhoea so that we will fertilise his field.") that he pleaded with me not to ask him to continue doing this.

His willingness to live in the village, get his hands dirty and eat the local food made him a model extension agent. Rural development is, at its heart, about people and relationships. Money, technology and expertise are useful but not paramount. Achieving sustainable and enduring transformation requires friendship, empathy, trust, humility and a willingness to be a co-learner in the adventure of change.

Liz's work

While I was busy working with the extension agents and visiting villages, Liz had a full schedule of her own. Early in our time in Niger, Liz was able to visit the villages, but she soon found that her time was fully occupied with family responsibilities, correspondence, keeping the project books, teaching literacy, meeting with local women's groups and interacting with those who lived nearby or came to our door. Working with women in rural villages was challenging. Very few Nigerien women were educated or worked outside the home. The days of rural women were filled with hard and repetitive physical labour and developing relationships required regular visits and ongoing commitment. One was heard to say, "Western women are not women at all! They drive cars, they have important jobs and they speak back to their husbands!"

Although family and other responsibilities made it challenging for Liz to travel to the bush regularly, she found many ways to contribute. Over time, Liz found that extension agents and farmers would come to her with their own questions or questions from their wives about health, vaccinations and other issues.

Her natural skill in caring for animals led to various small livestock-related activities including promoting improved methods of raising guinea fowl, ducks and rabbits to enhance nutrition and income. Some initiatives were successful and some not. When Liz started working with neighbouring families to increase the survival rate of ducklings, duck flocks multiplied rapidly. However, this endeavour came to a sudden halt when we realised that the ducks were not only digging up all the peanuts that neighbours had planted but were also

fouling people's compounds and taking food from the bowls of small children who sat on the ground to eat.

Every November, Newcastle Disease killed 90% of all chickens. The government had started a "barefoot" vet program in which young men were taught to vaccinate chickens against Newcastle Disease but there was limited uptake. Each year as November approached, Liz talked to our project extension agents and visiting farmers to explain how and why chickens should be vaccinated. The arithmetic quickly made sense to many: if they sold one chicken, they could pay for all the rest of their chickens to be vaccinated. However, the money was not the only challenge. At night most village chickens roost in trees or on the top of huts or fences. The whole village would spend the night before the vaccinations trying to catch their chickens in the dark. If the barefoot vet did not turn up in the morning, it was difficult to persuade the villagers to repeat the exercise for another day. Nevertheless, there was increasing acceptance of the value of vaccination.

Yardiya

Over the years, Liz employed various women to help in our home. She didn't always enjoy their constant presence in the house and the pressure of keeping them occupied, but, without help, all her time would have been taken up with cleaning and cooking.

From our second house help, Yardiya (Little daughter), we learned much about the realities of Hausa life and culture. At nine years of age, Yardiya was taken from Niger to Nigeria to marry a man who was 75 years old! She was his fourth wife and had to work all day for the other three wives. As soon as preparation of the evening meal began, she would run and hide in the bush. She did this for several years before her husband called a stop to it. By the time she came to work for us, her husband was reputed to be 113 years old and lived in our neighbourhood. The other wives had left, and Yardiya was caring for him alone. She worked for us during the day and nursed him by night. Yardiya was a diligent and reliable worker. In addition to cleaning, washing and cooking she sometimes looked after our children and helped me with my projects. When I introduced acacias with edible seeds, Yardiya winnowed and cleaned the seeds and experimented with recipes.

Yardiya taught us much about the realities of Hausa life and culture. At nine years of age she was married to a 75-year-old man who already had three wives.

Yardiya also worked as a village midwife. As part of the government program for local midwives, she could be picked up at any time of day or night by an ambulance. Sometimes Yardiya and other village midwives would be taken to see and learn from unusual deliveries. One night, she was picked up to watch a baby die of tetanus. The midwife had not used a clean razor blade to cut the umbilical cord and the medical staff wanted to drive home the importance of following instructions. Yardiya suffered from stomach ulcers all her adult life. Sadly, a ruptured ulcer took her life during our last term in Niger.

Liz's role in supporting me—providing advice, balancing the project financial records, hosting the myriad visitors who came to see me, running the Maradi guest house, corresponding with family and supporters, occasional home schooling, and raising a family besides—was indispensable to the success of our work. Many times, she put the wheels back on my emotional cart following criticism, mistakes or failure.

Health care

There were no paved footpaths or lawns, only dirt paths and open spaces. Open defecation was the norm on the outskirts of villages and even in Maradi itself. Animals lived in close proximity to where children played and food was prepared. Water was not plentiful. Windblown dust penetrated and covered every surface, and there was constant contact with the ground. Under these conditions a simple scratch could easily get infected and, if untreated, lead to serious illness and even death. Eye infections, diarrhoeal diseases, colds, coughs, meningitis, hepatitis, tuberculosis and sexually transmitted diseases were common. While some traditional treatments may have been helpful, concoctions made from the bark of trees could ruin livers, and poultices containing a mix of manure and herbs could fan the flames of infection. During our first four-year term our health was sorely tested.

Malaria was prevalent and despite taking chloroquine weekly, we occasionally got fever. Fortunately, in those earlier years, it was relatively easy to treat. We don't know how we contracted hepatitis, but it shouldn't have been a surprise given the lack of sewerage, the strong, desiccating winds which whipped up dust, and the frequency of shaking hands when greeting. To recover we were evacuated to Miango, on the cooler Jos Plateau in Nigeria. SIM had built a convalescent centre in the grounds of a guest house where missionaries took their holidays. It was quite traumatic for us when, upon arrival, Ben was whisked back to Niger to be cared for by another family. Ben showed no signs of infection, but children often show no symptoms and the guest house manager did not want to risk Ben infecting other guests. It was strange being separated, and we often imagined that we could hear him. When we were reunited after three weeks, it tore at our hearts that for a few minutes he wasn't quite sure if he recognised us.

Meningitis was more harrowing and potentially more deadly. For tea one Saturday night we ate pancakes. Around bedtime, I was getting a headache but was called out to take a villager (who was later diagnosed with meningitis) to the clinic. As the night wore on, I began to vomit, had fever, a worsening headache and my neck began to stiffen. The next morning, I was taken to the same clinic where a German doctor made the diagnosis. John Ockers, a pioneering missionary who had conceived of and built the Farm School from scratch, drove Liz and me the three hours to the SIM hospital in Galmi. I was lying in the back of John's station wagon. As it became harder to push my head back and my headache became severe, the seriousness of my predicament slowly dawned on me. I remember asking God not to take me; I felt I had not yet accomplished anything with my life. Doctor Ceton, a SIM medical missionary, did a spinal tap and was able to confirm that I had spinal meningitis. At first, I didn't mind the nurses sticking

Living in another culture broadened our appreciation of alternative child-rearing practices.

needles into me as it took my mind off the headache. But by the second day and as my headache subsided, my bottom felt like a pin cushion and I begged to be given oral antibiotics. To this day, the association of pancakes—which had nothing to do with contracting meningitis—and feeling very ill remains vivid, and I avoid eating them.

Though vaccination programs have improved the situation markedly, meningitis continues to claim the lives of thousands of people annually in the region. Several years later, both Liz and our daughter Melissa contracted the disease. To see a loved one suffer was more devastating than to experience it myself. However, our "suffering" was nothing compared to the heartbreak experienced by many Nigeriens and some missionaries who lost children or spouses, or who left the field permanently with broken health.

Open flames and kerosene lamps were a daily reality and so burns were common and often severe. I knew people who had emerged from burning grass huts looking like a sizzling roast. Some survived but were disfigured; others were not so lucky. Women and children who cook with open fires were particularly vulnerable. More than one Nigerien baby, snuggly tied to its mother's back, had been burnt when their mother's loose-hanging, wrap-around skirt caught fire.

I had my own close shave with fire. I was visiting a village and had just lit my gas lantern. It was the kind with a disposable canister, so when it ran out of gas, I quickly unscrewed the old canister and screwed on a new one. As the full canister is screwed onto the lamp, it is pierced by a spike and in the second or two it takes to screw it on tightly and form a seal a small amount of gas is released. In that fateful two seconds, the still-hot metal of the lamp ignited the escaping gas. Flames shot up to the roof, singeing the hair on my arms, eyebrows and forehead!

Fortunately, the seal had formed so that only the escaped gas ignited, not the whole container.

Before I could react, my hosts dragged me out of the hut, and everyone began spitting furiously on my singed arms! In the absence of water, that was the best they could do to cool my skin. Everything in that grass hut was tinder dry and combustible. There were loose plastic bags under the bed. The home-made mattress consisted of sewn plastic superphosphate bags stuffed with dry grass. The walls and roof were made of thatched grass, now years old, and I know the flames reached the roof because spider webs were singed.

House fires were common in Niger. Flames could consume the dry grass huts in seconds and, with strong wind, a village in minutes. Knowing the danger, I marvelled at God's mercy in my minor injury and was reminded of the words, "When you pass through the waters, I will be with you; and when you pass through the rivers, they will not sweep over you. When you walk through the fire, you will not be burned; the flames will not set you ablaze."[14]

We found a way to help some people with burns. While on holidays we came across some aloe vera in Nigeria, brought it back to Niger and multiplied it. Cutting the leaf with a sterile knife provided a safe dressing that was soothing to the burn and high in vitamin E. Many parents who had brought children with burns came back excited that the burn had healed so quickly and cleanly. We sent them home with their own aloe vera plants and encouraged them to give the pups—new aloe vera plants which emerge from the base of the mother plant—to their friends and relatives.

Even though Liz had no formal medical training, for the first few years she inherited the role of dispensing medicine to Bible-school students and staff families. By studying the relevant pages of the Hesperian Health Guide *Where There Is No Doctor* and avidly taking in the advice of other missionaries, she got through with no major mishaps.

Liz tried to teach mothers how to do as much "barefoot" doctoring as possible: soaking infected wounds or mosquito bites in warm salty water often eliminated the need for antibiotics; wiping soap on infected mosquito bites stopped flies from landing on the sores and further infecting them; aloe vera relieved burns; and a homemade rehydration mix (two tablespoons of sugar plus a half-teaspoon of salt in one litre of water) for diarrhoea saved many young children and adults from needlessly purchasing medicine or sometimes even death.

We were amused that Nigeriens willingly drank the mixture and got better while expatriates, who didn't like the taste, waited until they could take "real" medicine.

Scorpions were prevalent. Over the years, a number of people came to us seeking relief from the excruciating pain of a scorpion sting. We didn't have any treatment to ease the pain. It was a case of *sai hankuri* (just patience), a common Hausa refrain. One victim begged, then insisted, that I use a car battery to give him an electric shock on the site of the sting. It was commonly believed that this would bring relief. Whether it did or not, I do not know. Perhaps the fear and pain of the shock gave the patient something else to focus on. As no member of our family was ever stung by a scorpion, I became complacent. Then one morning right at the end of our time in Niger, without thinking, I went to put my boots on as I did every morning. But on this occasion I was gripped by an impulse to shake the boots first. I'm glad I did. A scorpion tumbled to the ground!

During our time living at the Farm School I drove many village people to hospital. If someone was sick enough to ask to be taken to hospital, you knew that they were in bad shape. Generally, people would only go to hospital once the pain was unbearable and they had exhausted every other option. By then, for the most serious cases, it was too late for doctors to save them. The fact that so many people never came back from hospital reinforced the commonly held view that hospital was where you go to die. It also cost money. There were no banks in rural areas and very few people had or kept cash on hand. A trip to the doctor required the sale of grain, chickens or some other prized possession and would mean going without something in the future. And, too often, "bush" people were looked down on and made to feel stupid for not understanding a system of which they had no prior experience. No wonder a combination of fatalism—"If God wills it"—and patience—the pinnacle of culturally cherished character traits—would often override a decision to seek medical help.

A number of recovered leprosy patients came to Liz from time to time. We soon learnt why they liked to keep cats. A common characteristic of leprosy sufferers is the loss of feeling in their hands and feet. If a mouse or rat started chewing on their fingers or toes in the night, they would not feel it, so having a cat to keep

Unique to each tribe, facial markings are made when a child is eight days old.

vigil was a necessity rather than a luxury. One day, Hannatu came to Liz for help. She had badly burnt her finger; as she handled her cooking pot, she had not noticed it dangling in the boiling water. By the time Liz saw it, the flesh was gone, and the bone was sticking out from what was left of her finger. Hannatu wanted Liz to clip off the protruding bone. Even though she insisted it was an easy procedure that anyone could do, Liz said, "No way," and gave her money for a taxi ride to the Dandja Leprosy Hospital for appropriate care.

Millet heads are carried to market on the backs of camels and donkeys and sometimes on people's heads.

Water and electricity

Ben had several close calls. When we were installing lined wells in six villages, contractors made the perforated concrete inner rings on our compound. These rings fit inside an outer concrete casing which is cast in situ. The rings are lowered down to the water level and, as sand is excavated, they sink below the level of the outer sheath. Water trickles in through the perforations leaving silt and sand behind and the ring holds a reserve of water that can be drawn from. Because large quantities of water are needed for the cement work, the first ring cast is not perforated and is turned into a basin by cementing the bottom. One day I was in our yard with Ben. When I turned around, he was gone. He'd raced over to the open-topped ring and jumped in for a dip. He couldn't swim and when I found him, all I could see was the tip of his nose, barely poking above the water as he stood on his toes to keep breathing.

On several occasions poisonous snakes passed through or took up residence in our compound where Ben played. Fortunately he was never bitten.

We depended on a diesel generator for our four hours of electricity per night. It powered lights, fans and a pump which supplied drinking water from a nearby well. It was on its last legs, and much of my time was taken up trying to maintain and repair it. On 27 December 1983, I wrote home about some of our challenges.

> On our way home from a wedding, an electrical fault caused the motor to stop. Liz and the kids caught a bush taxi and didn't get home till 2:00 a.m. I slept under a millet stalk lean-to and waited for someone to come back and tow me home. I still haven't located the ault! This week I've had fevers and a cold. It's been a bit miserable and our colleagues on the station are away—so I still need to manage the staff. Three out of the four vehicles aren't working, as well as a brand new pump.

When my brother Peter, a very competent mechanic, visited us, he was shocked by a number of things—for example, seeing the road through the holes in the rusty floor of the taxi we travelled in, and my mechanic's "workshop" which comprised of a bare allotment and a small, lockable shed in one corner. All repairs were undertaken out in the open; parts were placed on mats on the ground where they were exposed to sand and grit. There was no block and tackle, so engines were manhandled in and out of vehicles by apprenticed boys. Genuine spare parts were hard to come by. I particularly needed to be cautious about spare parts coming from Nigeria as it was common for used parts to be cleaned, repackaged and sold as new.

As our generator aged, it became less and less reliable. There were few qualified technicians to call on, so when a visiting SIM missionary who was a diesel mechanic visited unannounced, I leapt into action. My whole focus was on making sure the mechanic had everything he needed to fix the generator. Spare parts list in hand, we jumped into the Peugeot 504 station wagon to drive to town. As I began to back the car, I thought I heard my name called. The strengthening wind carried the sound away, and I was determined to have the generator fixed, so I didn't stop—until I felt a subtle thud. My mind computed what may have happened. The call had come from our domestic helper who could see everything unfold from inside the house. Ben had been playing in the yard. Momentarily I thought the rear wheel must have rolled on top of him and I almost continued in reverse to release him. Fortunately, a stronger inner voice overruled. "STOP!" I jumped out and, with my black greasy hands, picked up a dazed little boy who had been so eager to go with his dad, but tripped at the last moment and had fallen directly behind the wheel. If I had proceeded in reverse, I would have crushed him. I raced him in to Liz who at first thought that my greasy handprints around his waist were tyre marks. We took him to the station nurse. No damage had been done, but from that day onwards, Ben had a very healthy respect for cars.

Some people thought that we were irresponsible raising our children in Africa. Some missionaries did lose children, a spouse or their health. We were there because we believed that was where God wanted us to be. We committed our children and ourselves to his care. Sadly, while Ben was having these brushes with death in Niger, children in my safe and civilised hometown, Myrtleford, died by drowning in a back-yard swimming pool, being crushed under a family car, and from snake bites. Life is unpredictable wherever you live. It is more important for us to be where we are meant to be than to live where it might appear safer.

The windmills at Maza Tsaye had been shipped out from the United States by John Ockers and were over 40 years old by the time we arrived. As with other spare parts, finding the leather washers which pulled the water up from the well was a challenge. I aborted my first attempt to ascend the eight-metre-tall windmill, thinking that I hadn't signed up for this. However, thirst overcame fear and I eventually summoned the courage to climb up and top up the oil in the nearly-dry gearbox.

Dry season

During the dry season, the days are overcast, not with cloud, but with Harmatan dust, super fine sand particles (0.5 to 10 microns) which paint the sky grey and work their way into hair, ears, cameras and houses. Northeasterly trade winds blowing off the Sahara Desert carry large amounts of this dust, and it blankets everything in fine grit from the end of November till mid-March.

The dust was a challenge for Liz. It only completely settled once the rains had set in and thoroughly wet the soil. Many of the houses had louvre windows which did not keep it out. Sometimes, we could set the table and by the time we ate, the whole table would be covered in dust, with clean circles under the plates and glasses. At times, we would set the table with the glasses and plates upside down and the cutlery underneath. After a big wind, it was not unusual to fill a 10-litre bucket with the dust we had swept up from the floor.

The first time we experienced a significant dust storm, I was about to head into town in the car. Ben (aged four) and Melissa (aged two) were playing with neighbours. Yardiya, called to Liz and pointed to the sky. Liz saw a huge, purple and red-grey, mushroom-shaped cloud moving swiftly towards the house. She wondered if the French had conducted a nuclear test! Yardiya told her that it was just a dust storm. Liz sent a child to bring Melissa back to the house and she and Yardiya raced around shutting windows. No sooner had they finished than it went completely dark. Liz could not even see her hands held up to her face. Someone pounded on the back door; they had brought Ben home. Ben was amazed and said in fluent Hausa, "Why did God cause this darkness to fall upon us? Tell him to take it away!" Panicking children were sometimes lost, injured or even killed when they ran into things or fell down wells. For some weeks the Muslim teachers had been predicting that the end of the world was coming, so many people thought that this was the end. Thieves in the market, however, took the opportunity to fill their pockets.

An enduring memory of our early days was setting up a fan and wetting the sheets several times before bedtime. The sheets would be dry within half an hour, but the evaporation cooled the mattress sufficiently to get to sleep. Just before going to bed, we would dip the sheets one last time. If you could get to sleep before the sheet dried, generally you would sleep for a number of hours before waking up thirsty. In our second term we were able to install an evaporative cooler and our sleep, along with the family's health, improved markedly. The children no longer woke up five or more times in the night thirsty. Having a full night's sleep greatly boosted our resilience.

Wet season

The rainy season was my favourite time of year. From May to June the humidity built up, becoming almost unbearable with the extreme heat of the dry season. There was a tension leading up to the onset of rains. Daily, one automatically scanned the heavens for signs of rain-bearing clouds. At a certain point the prevailing winds shifted from northeasterlies to southwesterlies, bringing engorged clouds. First came strong, ground-level winds carrying

With the first two or three rains, from nowhere, thousands of beautiful red velvet mites and desert toads appeared in the fields.

blinding dust and debris. Often there was wind and dust, but disappointingly, no rain. If rain did come, the first shower often arrived as a horizontal blast of mud and grit, at which point all pandemonium broke out. Adults, chickens, goats and sheep rushed to seek shelter. Children shrieked with delight as they ran wild in the rain to play. You could hear the wave of squealing as the rain moved across a village. With the first two or three rains, from nowhere, thousands of beautiful red velvet mites and desert toads appeared in the fields.

Advice in Abalak

As I grew in experience, I began to receive requests to make assessments and give advice on reforestation in other parts of Niger. One such invitation came from Jeffery Woodke,[15] who worked for Youth With A Mission (YWAM). I wanted my staff to get as much experience as possible. So, with two of my colleagues, I travelled from Maradi to Abalak, where Jeff worked. Normally I didn't mind the heat, but this day was extreme. It must have been close to 50 degrees Celsius. Even in sandals, the hot sand in the compound burnt the soles of our feet and the heat in the shade was intolerable.

We were greeted by a petite Swiss missionary. I marvelled at her endurance and energy. The missionary house was built from mud brick with a corrugated-iron roof and had no insulation or ceiling. Their project was helping the local community with relief food and agricultural development. They had constructed low, rock weirs across the dry stream beds so that when the floods came in the rainy season, water would be retained long enough to soak into the ground. Crops could then be successfully grown using the moisture stored in the soil. Jeff and his team were much loved and highly respected by the local community.

Jeff took us to visit a local Tuareg chief who treated us to the best desert hospitality: generous servings of camel meat and milk for breakfast, lunch and dinner. I am not a big meat eater but to refuse would have been a great offence. Drinking s*haye* (tea) with visitors is an obligatory ritual and gesture of welcome. We sat on woven mats placed on the white sand, sipping the sweet brew, and listened to the chief's stories. "When I built this house, I needed sand for the reception area. There was no sand locally so I travelled over 30 kilometres in my Land Rover to obtain nice, clean sand. If only I had waited! Today the winds have brought the sand to me." A dune had formed in his compound and the sand was pouring into his house through a window!

The chief lived in a very remote area. Curious, I asked him what he did when someone got sick.

"No problem," he replied, "We give them camel urine!"

Surprised, I persisted, "But what if they are really sick?"

"No problem; we prevent the camel from drinking and then we give them concentrated camel urine!"

The Tuaregs fiercely resisted French colonial rule but in the end their uprisings were no match for the more advanced weapons of French troops. After numerous massacres on both sides, the Tuareg were overpowered and obliged to sign treaties in Mali (1905) and Niger (1917). Nature dealt them an additional blow in the 1970s when they lost most of their camels and other livestock during an unprecedented drought. Once lords of the desert who controlled much of Niger, the Tuareg people have lost most of their herds and now face increasing poverty and desert encroachment. Mistrust of authority lingers to this day and in recent years their homelands have seen uprisings and incursion by extremist groups.

CHAPTER SIX
Sowing into the wind

"Hey, White Man, why don't you buy a car?" In her long life, the old woman had rarely seen a white man. She had certainly never seen one *walking* through her village. It was pre-rainy season, 1982. Moussa and I had loaded several hundred tree seedlings onto the Ford F100 pickup and trailer. We had departed the Farm School nursery early that morning to deliver the seedlings to the village of Gangara, on the outer limit of our project area. Three-quarters of the way into our journey, in the middle of nowhere, a sharp root protruding into the deep wheel ruts slashed not one, but two tyres. We only had one spare. Thus began our 23-kilometre trek back to the Farm School in the midday sun.

There I was, the resident expert, but what did I know about reversing land degradation and desertification in Niger? Even so, I put everything I had into doing so. I reasoned that if deforestation was one of the root causes of the problems people were facing, then reforestation must surely be part of the solution.

Visiting the villages around Maradi, I was shocked by the condition of the land. It was teetering on the brink of ecological collapse and barely able to support life. How could I, fresh out of university and with no experience, make any difference? Much of the woody vegetation that had been there only 20 or even 10 years before my arrival was gone, mostly cut down by farmers. During a series of droughts, people felled even the last trees to sell the wood to buy food. On average, only four trees per hectare were left. Today we know that, depending on the species regenerated and how farmers manage them, farmland can comfortably support over 100 trees per hectare and yield more grain and other benefits than land without the trees.

> Where there are no trees, we have to plant trees. It's self-evident. Right?

Without trees, the soil becomes exposed, eroded, compacted and infertile. Nothing grows on the resulting hard pans.

Value of trees

Trees and their timber and non-timber products are to the rural African economy what oil and its derivatives are to a modern industrial economy. Wood is used for fuel, fencing, house construction, furniture, tool handles, saddles, Muslim prayer boards, sieves, carts, ploughs, looms, the lining of traditional wells, the Y-shaped gantry, shaft and pulley that make it easier to draw water, bows and arrows, spears, musical instruments, goads, walking sticks, mortars and pestles for grinding grain and condiments, and on, and on. Charcoal is used by rural blacksmiths to melt aluminium for castings and in forges for making axes, hoes, knives and other utensils used in daily life. Trees also provide fodder, wild foods, traditional medicines and ecosystem services like pollination, climate regulation, pest control, soil retention and fertility, spiritual comfort and cultural meaning. The list could fill a book on its own.

Burden of tree loss

The burden from the loss of trees affects everyone, but none more so than women and children who are tasked with collecting fuel, fodder and wild foods. As the forest retreats from a village or town, women and children must walk further to bring back the same amount of wood. In Niger, it could mean walking two to three hours every second day to collect fuelwood. They are very vulnerable: in some places to attack by wild animals, in others to rape and abduction. The long trek in the hot sun carrying heavy loads takes a toll on health and energy levels. Time spent collecting wood is time not spent on education, economic activity or socialising. Eventually there is simply no wood to collect. Then, the negative impact of deforestation multiplies as substitutes are sought, such as straw and animal dung, two commodities that the soil desperately needs for protection and fertility. Infertile soils low in organic matter result in less robust crops, lower yields and lower livestock production, which contributes further to hunger and poverty.

Population pressure

At Niger's independence in 1960, the population of the former French colony was under 3.4 million. When I arrived in 1980, the population hadn't quite reached 6 million and by 1984 it was 6.72 million. The cities were growing rapidly and with them the demand for firewood. The population of the city of Maradi more than doubled in 10 years; from 45,000 inhabitants in 1977 to 110,000 in 1987. People in the countryside were poor with few means of earning income, so trees were cut down and sold to generate income. As rural populations grew, the cycle for slash-and-burn agriculture shortened. It reached a point where no land was being rested, except areas which had become so infertile that they were abandoned.

Effects of deforestation

Deforestation is a precursor of land degradation and desertification. It is estimated that 74% of rangelands[16] and 61% of rainfed[17] croplands in Africa's drier regions are damaged by moderate-to-severe desertification. This occurs on all continents except Antarctica and affects the livelihoods of millions of people, especially the poor in drylands. These drylands occupy 41% of the earth's land area and are home to more than two billion people—one-third of the human population in the year 2000. They include all terrestrial regions where water scarcity limits the production of crops, forage, wood and ecosystem maintenance. Land degradation has a direct impact on predisposing land to drought, flood, landslides and plagues of pests and disease.

> Desertification also affects global climate change through soil and vegetation losses. Dryland soils contain over one quarter of all organic carbon stores in the world as well as nearly all the inorganic carbon. It is estimated that 300 million tonnes of carbon are lost

While trees are often seen to be an impediment to cultivation, the benefits outweigh the disadvantages, and strategies to minimise negative impacts can be adopted.

to the atmosphere from drylands as a result of desertification each year—about 4% of the total global emissions from all sources combined.¹⁸

Deforestation directly and negatively affects the lives of millions of people. By contrast, reforestation can be a powerful tool to improve the lives of the most vulnerable people in the world. Reforestation can improve microclimatic conditions and soil fertility, reduce the incidence and severity of flood and the impact of drought, increase food security and water availability, build resilience to disasters, and encourage economic development and diversification of income streams.

I asked myself many times how huge parts of Niger could have been so devastated given the relatively small population. This is not a new phenomenon. It seems our predilection towards removing trees is universal. The US Farmers Register of August 1833 hints at both the rapacity of the destruction, and the predictable consequences: "They wage unmitigated war both against the forest and the soil—carrying destruction before them and leaving poverty behind." The sheer extent and severity of the destruction in the 1800s is staggering considering that there were only around one billion people living on earth at that time and they had no chainsaws or bulldozers. In his seminal work, *Man and Nature: Or, Physical Geography as Modified by Human Action* (1864), George Perkins Marsh, a politician who became a US diplomat to Europe, described a situation that makes one wonder if there is anything new under the sun:

> Territory larger than all Europe ... has been entirely withdrawn from human use, or, at best, is thinly inhabited ... There are parts ... where the operation of causes set in action by man has brought the face of the earth to a desolation almost as complete as that of the moon; though ... they are known to have been covered with luxuriant woods, verdant pastures, and fertile meadows.[19]

No doubt, my hero, Richard St Barbe Baker would concur.

Trees are weeds

In addition, during the colonial era, Europeans brought the concept of monoculture to Africa. If you want to produce large quantities efficiently you provide ideal conditions for the crop of choice. The crop favoured by the colonial administration was peanuts, and Maradi had become a major production and processing centre. To boost peanut production, extension programs promoted cultivation with donkey and ox-drawn ploughs and the use of peanut planting machines. Farmers were told that "modern farmers" cleared trees and stumps from their land to aid cultivation and to increase peanut yields. A big peanut oil factory remains to this day, but it stands idle for most of the year.

Most farmers in Niger viewed trees on farmland as "weeds" which needed to be eliminated because it was believed that they competed with food crops for nutrition. Even the language appears to devalue trees. The Hausa word for tree, *ice* (pronounced "itchy") is the same word used for firewood. Trees were also believed to attract grain-eating birds—something that no farmer, especially one living on the edge of hunger, wants to see. Over recent decades, farmers who could afford to buy ox-drawn ploughs removed all obstacles such as trees and tree stumps so that the oxen could plough unhindered. Even farmers who cultivated their fields with a hand hoe cleared any vegetation before sowing their crops.

Years later, while I was visiting Senegal's peanut basin, a farmer told me that he remembers clearly the day his life changed forever. At a meeting in the village square an extension agent demonstrated planters and ploughs which were being provided on credit. It was his job and that of the other village boys to clear the land and remove stumps and roots from the field. The experience of every boy growing up in Niger's peanut growing districts would have been similar. In the 1970s, the Maradi Productivity Project did the same thing thanks to an intense campaign to "modernise" agriculture, which involved the removal of trees. Those not practising modern agriculture—clearing trees and using animal traction equipment—were shamed. Poor farmers seeking a way out of poverty found the offer irresistible.

The combined effect of these practices was the almost complete destruction of trees and shrubs in the agricultural zone of Niger between the 1950s and 1980s. This destruction had devastating consequences. Deforestation worsened the adverse effects of recurring drought, strong winds, high temperatures, infertile soils, and pests and diseases on crops and livestock. Combined with rapid population growth and poverty, these problems contributed to chronic hunger and periodic famine. The severe famine of the mid-1970s provoked a global response and, finally, stopping desertification became a top priority for governments, donors and NGOs.

A daunting task

The average Nigerien farmer is almost totally reliant on the success of the annual millet crop in an environment which conspires against it two years out of every three. Under the land management methods of the time, hunger was less an unforeseen calamity than something which could be easily predicted.

With these things on my mind, I read, "His divine power has granted us all things that pertain to life and godliness, through the knowledge of him who called us to his own glory and excellence."[20] Living in Niger, it all seemed a bit fanciful. Maybe it was true in temperate countries that God had given everything needed for life, but here, in Niger?

Some fields had become so degraded that the erosive forces of wind and water created "hardpan" soil, compacted and wind-polished earth, set hard like concrete. Areas affected in this way were like parking lots which supported no green plants at all for decades. One day while walking across a hard pan surface following rains, looking down I saw seeds germinating through some cracks. It was astonishing. That morning during my devotions I had read, "When you send your spirit, they are created, and you renew the face of the ground."[21] God was not only in the business of creation and of ministering to human spiritual and physical needs;

he was concerned with renewing the face of the earth. The longer I lived in Niger, the more I learnt about the vegetation and resilience of nature. I realised that all was not lost. I was not in this fight alone. With baby steps, I began simply asking God for wisdom and insight.

Community nurseries

Conventional approaches to reforestation are expensive and labour-intensive. They faced insurmountable problems. I did everything that I knew how to: I read widely, and I consulted experienced people in the field and visited their projects. I experimented with different techniques. We planted exotic species which was standard practice at the time because indigenous tree species were perceived to be of low economic value and to be slow growing. We planted indigenous species. Still trees died. Every year Niger celebrated *trois août* (3 August), as national tree-planting day. This made absolutely no sense to me. In a good year, Niger only had a four-month rainy season. Why would you wait till the month before rains stopped to plant trees and then have them face an eight-month dry season? So, we started planting with the first decent rains, and, with those willing to do the work, even before the rains to give the trees a head start. We had a central nursery and had to transport the trees long distances to the villages on rough roads. This was expensive and some trees were damaged in the process, so we encouraged community nurseries. The concept of "community" nurseries sounds wonderful, but few people in the community buy into the idea and all the work lands on one or two keen individuals. When the hard work is done though, everyone in the community wants their share of trees—for free. I tried promoting nurseries owned by individuals. These were generally much better run, but few people wanted to do the work or pay for the trees.

In any case, success in raising trees in a nursery, where you think you have control, required constant vigilance. In the height of the dry season, nurseries are an oasis of green surrounded by a brown parched landscape. In this barren landscape, nurseries attract a whole zoo of damaging creatures including lizards which bite off the leaves and growing tips as they appear, birds which scratch out the emerging seedlings as they search for insects, toads which kill germinating seedlings by squatting on the cool moist surface of the planting bags, grasshoppers and caterpillars which strip the leaves off the seedlings, and sometimes termites, which even eat the black polythene planting pots, causing the soil to fall away from the roots when you lift the seedling up for transplanting. Nurseries were always placed near the village well, because in the high dry season temperatures, seedlings needed to be irrigated one or two times per day. However, the well is also where all the village livestock congregate as they wait to be watered. Is there a more able herbivore than the humble goat, with time to spare, an empty stomach and the ability to penetrate the most intricately woven thorn fence? It's all very well to want reforestation to be sustainable and replicable to the communities involved, but some things such as insisting on building thorn and brush fences just aren't practical. On the other hand, supplying wire fencing, which afforded better protection from goats, had its limitations. Wire, a rare and expensive commodity, is easily stolen at night by somebody who thinks they have a better use for it, such as meeting their own fencing needs, or manufacturing toy wire-frame jeeps which tourists love to buy.

Problems in the nursery were only the beginning. Survival of transplanted trees was greatest when seedlings were planted immediately after heavy rain—the exact time when any farmer worth their salt will be out planting their own field crops. This made it difficult to get enough labour for the task, particularly in years of food shortage. It was ideal if newly planted seedlings could be watered in, but this was a luxury. Should the seedlings survive planting, they could be subject to strong winds and sandstorms which could strip the seedlings bare of their leaves, sand blast them and even bury them altogether. Pests, competition from weeds, drought and browsing by hungry livestock took their toll and negated our efforts.

Well depths ranged from 40 to 100 metres. It is back-breaking, time-consuming work to draw water daily for any sizeable nursery and only the fully committed will continue to do this on a volunteer basis. I cajoled villagers, and I rolled up my sleeves and worked with them. At times, I even used the blunt tool of incentives, usually in the form of tools or prizes for good work, to try to get people to grow trees for long enough to see the benefits for themselves. Then, hopefully they would continue planting trees without me. However, incentives are a poor substitute for walking alongside people on a journey of learning, discovery and adoption of new ideas based on knowledge and understanding. One time, on the advice of my assistant and against my better judgment, I loaned a chief some money to purchase oxen. Of course, he

When agriculture fails to meet the needs of farming families, they often cut trees to earn income by selling firewood. This further exacerbates the vicious cycle of destruction and poverty.

never repaid the loan, nor did his performance on tree restoration improve.

There are any number of development projects in which activities stop once the last dollar is spent. Despite all my efforts, nothing worked in a cost-effective, sustainable way. In fact, to many, I was the "mad white farmer" and my ideas were just silly.

Counterproductive policies

Well-intentioned but misguided laws were another obstacle to reforestation. In Niger, the government automatically adopted the French forestry code put in place by the colonists. All naturally occurring trees, even those growing on privately owned farmland, belonged to the government and were protected by law.

These laws placed the responsibility for managing trees in the hands of officials who drew a salary regardless of whether or not trees remained in the landscape. These authorities were sometimes corrupt, usually remote and often simply too under-resourced to be able to police the landscape effectively. Because foresters could not be everywhere at once, people cut down trees whenever and wherever they could.

To clear trees or even harvest their branches, individuals were required to buy a permit. The hassle of finding the money and getting to town to obtain a permit was a big disincentive to follow the law. There was also a very real possibility of being fined if some one else stole or harvested trees on your land. Even landowners inclined to leave trees on their land believed that it was better for them to benefit from the destruction or removal of trees before someone else did.

In time, almost all of the southern third of Niger became deforested. Much of the destruction can be linked to counterproductive policies and laws. They were created to protect trees but they had the opposite effect; they accelerated the removal of trees!

Correct assumptions make all the difference to how you define and tackle a problem and your chances of success. There is certainly a place for laws protecting trees but they need to take into account the desperate needs of millions of smallholder farmers and their families who depend on the natural resource base, including trees, for their livelihood. Those who depend on trees are the people best placed to manage trees sustainably. They should be rewarded not punished for looking after trees. Farmers who benefit from trees will care for them.

Opposition from farmers

However, the biggest impediment was the opposition and derision from farmers. Such opposition put a very big damper on any enthusiasm somebody already combating the elements might have. Some community members saw trees as a threat to their livelihoods. They believed that the planted area would be off limits for grazing, further limiting their already restricted access to fodder. Others opposed tree planting because it would be going against tradition. Many trees were simply pulled out when nobody was looking. Poor community ownership and the lack of individual or village level replicability meant that no spontaneous, indigenous revegetation movement arose out of these intense efforts.

Nothing worked in a sustainable, financially viable, satisfying way. Our best efforts seemed to be just a big waste of time and money. Fortunately, I had the full support and understanding of Liz who believed in me even when I didn't believe in myself.

Worse still, the people we were trying to help didn't seem to care. "After all," they reasoned, "who in their right mind would plant trees on valuable arable land? And couldn't Tony see that what we really wanted was food and cash crops?" My efforts seemed as useless as trying to push back the sands of the Sahara Desert with a hand broom and shovel. With the conventional reforestation methods, which I was using, I knew that I would never have the significant impact necessary to solve the enormous problem of desertification. I lost count of the number of times I travelled through nearly treeless plains thinking it was hopeless.

What happened in Niger in the 1970s and 1980s (left) mirrored the American dust bowl experience of the 1930s (right).

Despite the high cost and failure rate, project after project, including our own, continued to raise and plant seedlings year after year. Doing the same thing over and over again and expecting different results is insane. Were we all mad, or were we simply so busy solving the problem, we didn't stop to analyse what was self-evident? I did analyse it, repeatedly. I just didn't have any alternative. Even though ineffective and uneconomic, reforestation through conventional tree planting seemed to be the only way to address desertification. Despite predictable failure, the same approach was repeated each year. We did what we thought was best but we failed to test our assumptions. Today, from India to Nigeria, countries are still trying to outdo each other by breaking the record for the greatest number of trees planted in a single day. If only they would set out to break the record for the greatest number of trees surviving after one year! And of course, learn from the Niger experience.

In *The Dust Bowl*, Ken Burns' documentary about the American mid-west, the farmers he interviewed recounted their experiences:

> You kept thinking that tomorrow things will change, so you kept doing what you were
> doing. Plant over and over again, hoping that this thing was over, and you were going

103

to have a crop. They couldn't live without hoping that things were going to change for the better. (Clarence Beck)

We learn slowly, and what didn't work, you tried it harder the next time. You didn't try something different; you just tried it harder, the same thing that didn't work. (Wayne Lewis)[22]

To me, this dust bowl experience describes perfectly what is being repeated to this day in many well-intentioned reforestation programs globally.

The trees we planted were dying because of neglect, animals, drought, sandstorms or termites. However, the principal cause of failure was the indifference and even open hostility of many people to the idea of reforestation. It seemed incredible to me at least, that some people actively opposed tree planting. Some thought that we were putting ourselves above God by planting trees: "It is God's work, and if he wanted trees, he would plant them!" Others thought that having trees on their land would just bring disputes, either with the authorities, or with people caught stealing them. In fact, the lack of private tree ownership and laws and enforcement mechanisms to discourage theft gave scant incentive for anybody to bother with tree-growing at all. Additionally, it was "common wisdom" that trees could not give a return to their owner in his lifetime because they grew too slowly, and that competition and shading from trees lowered crop yields. The problem with common wisdom is that whether it is right or wrong, it is impervious to any ideas that challenge it.

Project failure a typical and universal experience

It wasn't that I was particularly inept. In some African countries deforestation rates exceed planting rates by 3,000%.[23] In Niger, despite millions of dollars spent on nurseries and planting, weeding, fencing and guarding seedlings, few projects made any lasting impression. It is estimated that some 60 million trees were planted in a 12-year period and that less than 20% survived.[24] Having seen much of the country, even 20% seems generous. A separate study conducted in three West African countries found that tree planting success is often limited in size and scope. In an Earthscan paper it was reported that there was "disappointing" progress in village woodlots in the Sahel where, between 1975 and 1982, $160 million was spent on various community forestry programs. By 1982 the achievements were about 20,000 hectares of "not doing very well" plantations at a cost of approximately $8,000 per hectare.[25] At 305 million hectares, it would cost $2,400 billion to reforest the Sahel! Clearly, at that cost and with these methods that is not going to happen.

God forgive us for destroying your creation

After two and a half years of mounting frustration I felt like a failure and was ready to give up and go home. But deep down, I knew that I was meant to be there and had to keep trying. Again, I reached out for help. I asked God to forgive *us* for destroying the gift of his beautiful creation. While at that time I did not think I had directly contributed to deforestation in the Sahel, my Western lifestyle, which involved rapacious over-consumption and waste of natural resources, pollution and destruction of habitats, had contributed to a global problem. Because of this destruction, people were suffering; they were hungry, impoverished and fearful of the future. I asked God to open my eyes and to show me what to do.

CHAPTER SEVEN
The forest underground

One day in May 1983, my vehicle was pulling a trailer loaded with tree seedlings. I stopped to reduce the air pressure in the tyres to enable the vehicle to travel more easily over the loose sand. Usually, I was in a hurry to get on with the job of planting trees. But not this day. Scanning the barren landscape, the futility and hopelessness of it all weighed heavily on me. North, south, east, west; as far as I could see there were empty, windswept plains. Even if I had hundreds of staff, a multimillion-dollar budget and many years to do the work, using the methods I was currently using, I would never make a significant or lasting impact.

As I was about to get back into the vehicle, a bush on the side of the road caught my attention. Thinking that the bushes scattered across the landscape were just shrubs or weeds, I had never given them a second thought. I walked over and took a closer look. Combing my hand through the foliage, I let the leaves slide between my fingers. Tree leaves are like a species signature, and these leaves had the distinctive double-lobed shape of a camel's hoof print. *Piliostigma reticulatum!* The Hausa call it kalgo. This was not a bush. These leaves belonged to a tree. It had been cut down, and I was looking at shoots sprouting from a stump. These "bushes" were, in fact, dormant trees—ready and waiting to recolonise the land. I was surrounded by trees. I was standing on a subterranean forest.

> What difference would a few seedlings make in this landscape? I was overwhelmed by the futility of it all.

In that instant everything changed. This was the answer to my prayer. This was the solution I had been seeking. And all the time it had literally been under my feet! Or as the Hausa would say: "The chicken went to sleep hungry. The next morning it was surprised and annoyed to discover that it was roosting on a granary."

The tree-stump thing

When a tree is felled, for most species, much of the root mass remains alive. Because it has access to soil moisture and nutrients and a large store of starch in the roots, the tree has the capacity to regrow rapidly from the stump. Under the prevailing traditional agricultural practices, each year re-sprouting stems would grow to about a metre in height. Then, in preparation for sowing crops, farmers would slash the regrowth and either burn the stems, branches and leaves to fertilise the soil, or take them home to use as fuel and fodder for animals. For as long as this regular slashing and burning continued, these "bushes" would never grow into trees, and the forest would remain underground.

I knew that there were millions of such "bushes" dotting the landscape, constituting a vast, underground forest. This was a game changer. Instead of planting trees we just needed to allow the existing trees to grow. We began experimenting immediately. By selecting and pruning the strongest stems and culling surplus stems, we were rewarded with rapid growth and superior form. With a new lease of life, I told my extension agents to try this on their own land and find individuals in their village who would be willing to follow their example.

Getting down to business

In 1983, our project staff worked in a dozen village communities. The villages varied in size from just a handful of households to over a 100 family compounds. Our team taught farmers how to select, prune and manage these "bushes" to recreate agroforestry parklands. Rather than being based on tree-planting, this system used what was already in the ground. Rather than relying on outside expertise and resources, it was owned and driven by farmers and communities.

I conducted village meetings and discussed with village chiefs what this discovery could mean for them and their communities. However, in 1983, I could only convince one or two farmers from each of the initial 12 participating villages to try out this idea on a small corner of their land. The early adopters included Hashimou's uncle. They were perhaps immune to criticism or felt sufficiently financially secure to take a risk. They became the butt of many jokes

1. Select species and stumps or trees

Step 1. Survey land for sprouting stumps or seedlings and identify what species of trees are present.

Step 2. Select the species and stumps to be regenerated.

2. Prune and manage

Step 3. For each stump, select three to five stems to keep and prune away the unwanted stems.

Step 4. For each remaining stem, prune off side branches up to halfway up the trunk.

Step 5. Protect the stems while they are growing.

3. Maintain and utilise

Step 6. Prune unwanted emerging shoots every two to six months as needed.

Step 7. Utilise tree for planned purposes: harvesting branches, portions of wood or the whole tree as necessary.

If we could look beneath the surface of the ground, we would see that for most species, when a tree is cut down, 30 to 50% of the mass of the tree remains—alive and able to resprout. This partly explains why FMNR-regenerated trees can grow so quickly. In many cases they are supported by a mature root system containing stored starch and with access to soil nutrients and water, normally beyond the reach of transplanted tree seedlings with their small root mass.

and were ridiculed and mocked. They were accused of being lazy for not clearing their fields and foolish for leaving trees on their crop land.

I asked these innovators to select the stumps they wanted to revive and decide how many shoots they would allow to grow on each stump. Fearing that the trees would compete with their crops, initially they were very cautious and left only a single stem to grow from each selected stump. Excess shoots were cut and side branches trimmed to halfway up the stems. I encouraged them to return regularly to prune new suckers and branches. I made no hard and fast rules. Farmers were given guidelines but were free to choose which species they managed, the number of stumps per hectare they cultivated, the number of shoots per stump they left, the method of pruning and the time span between subsequent pruning and harvesting of stems. I had learnt from Roland Bunch that, to succeed, this method had to be farmer managed, not Tony managed. Through trial and observation we learnt from each other and progressively modified our technique accordingly.

After just a few months, it was obvious that this technique was working. It was cheaper, easier and more successful than tree planting. Despite all the effort, most transplanted seedlings died. With this method there was a near 100% survival rate! That was, until others began to interfere. As there was no other firewood available for cooking, some people cut and stole the young trees out of sheer desperation. Wood was a scarce and valuable commodity and, traditionally, trees were considered a gift of God for all to exploit. Others stole the trees out of greed or even to settle old scores. Yet others opposed the work because it went against tradition.

Initially we had no name for this approach to reforestation. In a stroke of pure genius, Moussa came up with the term *Sassabin Zamani*. In the Hausa language, *Sassabe* is the term for traditional land clearing that involves the removal of all bushes, weeds and crop residues. *Zamani* is the word for "this generation." Thus, *Sassabin Zamani* (modern land preparation) was conceived, and the term is still used universally amongst Hausa speakers to this day.

FMNR is born

In 1990, Barry Rands, a good friend from the United States Agency for International Development (USAID), visited our project. Barry had with him a camcorder—a fairly new device at the time—and filmed interviews with farmers in their fields.[26] It was Barry who coined the term "Farmer Managed Natural Regeneration" (FMNR). The name captured both the technique being used and the actor responsible for using it. The inspiration for the term came from the earlier work of a mutual friend, George Taylor II, in developing Farmer-Managed Irrigation Systems (FMIS) in Nepal.

We see what we believe

This method of managing trees was not new. It is a form of coppicing, pollarding and pruning which has a history of over 1000 years in Europe. Examples can be found in various parts of the world. It was new, however, to many farmers in Niger who, since colonisation, viewed trees on farmland as "weeds" that needed to be eliminated.

Why hadn't I seen the bushes for what they were before? Why had much smarter, more highly qualified and better-paid people also missed seeing them? Growing up in Australia I had seen Eucalyptus trees sprout profusely after being cut, yet I did not apply that knowledge to what I believed at the time were useless bushes on farmland. I had not been looking for a

Norbert Akolbila, the founder of Movement for Natural Regeneration (MONAR) in Ghana, demonstrates how to select and prune stems growing from tree stumps.

solution in the ground. I had been looking for a solution among conventional reforestation strategies. I had been stuck in a groove or habit of thinking and assumed that there were no trees left in the fields and that the solution could only involve raising and planting seedlings. We do not believe what we see. We see what we believe.[27]

FMNR and stakeholders

Land needing reforestation is usually inhabited or at least used by people. In such settings, it is very important to engage all stakeholders: sedentary farmers and nomadic herders who pass through only for a few months of the year, men, women and children, government representatives including forestry and department of agriculture agents, merchants and staff of other NGOs, traditional chiefs, religious leaders and others. Great effort is required to ensure that they understand the benefits to them and are not only in agreement but want the land reforested. We build their capacity and equip and empower them to make decisions, enabling them to benefit from their efforts. These activities are critical elements in an FMNR project. The success of FMNR lies in working with communities to build an enabling environment that allows them to develop their capabilities in organisation, networking, dealing with authorities and creating governance systems.

I did not have this knowledge at first. The Nigerien government model was to employ a *vulgarisateur* (populariser) and an *animateur* (host, facilitator). Too often this seemed to rely on hubris and showmanship. Few were convinced by their performances and, generally, results were not long lasting. In those early years I followed my father's example; I showed genuine concern, tried to help people at their point of need and worked most closely with the friends I had made in the process. It was these insider contacts who eventually convinced others. It was only decades later when I joined World Vision that I was to learn from my colleagues about more formal approaches to participatory development practices and stakeholder engagement.

Unnatural divisions

The artificial division between the disciplines of forestry and agriculture presented another obstacle. In reality, there is no hard and fast cut-off point where pure agriculture and pure forestry begin or end: they are two components of a continuum. These constructs create their own methodologies, idealised outcomes and bureaucracies. At one extreme, agriculture cannot entertain the presence of trees on cultivated land and at the other, forest departments set up primarily for the protection of trees look unfavourably on the harvesting of trees or the cultivation of crops within the forest. Historically in Niger, foresters had legal jurisdiction over trees on agricultural land and, following national policies, they had been tasked to protect them and prevent their disappearance. Agriculturalists, on the other hand, had no incentive to promote their protection.

Where people are the problem, people are also the solution

I reasoned that if I could just convince individuals and communities that it was in their best interests to allow some of these "bushes" to regrow into trees, then the rest would be relatively easy. After all, everything you needed for reforestation was already in the ground! No nurseries, no transport, no digging and planting and much less protection was required. And, indeed, that was the start of a massive landscape transformation which has since swept across the

nation of Niger, and today it is sweeping across many other nations around the world. But convincing people to change was not as easy as I thought it would be.

Though FMNR is easy to learn and do, this discovery was revolutionary. The experts had spent decades implementing futile anti-desertification measures and failed miserably. By the 1980s, donors who had spent tens of millions to "push back the sands of the Sahara" had stopped giving to tree-planting schemes. Despite universal appeal, good press and political point-scoring, tree planting had failed. In "discovering" the underground forest, the battle lines were immediately redrawn. As it turned out, I was not fighting the Sahara Desert. It was not so much a question of how big my budget was or which technology could be used to solve the problem. This was primarily a battle over false beliefs, negative attitudes and destructive practices towards trees and the environment. This destructive trio would need to be challenged patiently, respectfully and persistently, and truths and positive attitudes would need to be credibly presented as alternatives. If it were people's destructive actions that had brought the environment to its knees, it would require people's restorative actions to reverse the damage. If people are the problem, they must be part of the solution.

Assumptions determine everything

The American writer and professor of biochemistry Isaac Asimov wrote, "Your assumptions are your windows on the world. Scrub them off every once in a while, or the light won't come in." If your assumptions are wrong, whatever you do will not give the most cost-efficient or best result. You will be blind to solutions that are literally at your feet, as you may not be addressing the root causes of the problem. You may even do things that are totally unnecessary. Like those before me, I assumed that the trees had been completely removed and that tree planting was the only way to restore them. Hence raising and planting seedlings was necessary. I had heard and also assumed from my short observations that indigenous trees grew too slowly to meet the ever-growing demand

> "I thought that perhaps my grandchildren might enjoy the benefits of the trees I nurtured but after only one year my trees are protecting my soil and crops and giving me fodder and wood for fuel."

for fuelwood. Grown from seed, indeed, indigenous trees grew very slowly compared to exotic neem and eucalyptus trees. Foresters, because of their training and experience planting fast-growing, uniform trees for timber production, tend to be biased against indigenous trees, which often have crooked trunks and branches. And so, in Niger as in many other countries, exotic species were planted. Many believed that "ignorant" farmers were the main problem because they were the ones who had cut down the trees in the first place. As it turned out, those assumptions were completely wrong.

Going native

Everything you need for reforestation already exists in the landscape. Indigenous trees are far from useless, and "ignorant" farmers are essential for achieving large-scale reforestation.

Unlike many exotic trees, indigenous species are adapted to local conditions. While it was true that most indigenous trees in Niger would not produce high-value millable timber, they are extremely useful for fuel, traditional construction, fodder, medicine, dyes, rope, fertiliser, shade and reducing wind speeds and high temperatures, building soil fertility and more. Historically, indigenous forests were actually the supermarkets and hardware stores, the chemists, the farm supplies depots, water regulation and purification units, the worship and cultural centres and much more. Besides, indigenous trees regrowing from tree stumps can grow amazingly quickly even in dry environments.

Walking the talk

My local staff were the front-line troops of the movement. Their hardships are unheralded—they drank the same well water and ate the same food as the villagers and sometimes suffered from the same illnesses as the people they served. They lived in grass huts and mud-brick houses without electricity and endured both high and low temperatures and sandstorms without complaining. They were often the butt of jokes, regularly ignored and sometimes targeted with physical or emotional harm or damage to their

> "No matter how well we cared for them, the trees we planted died but with FMNR we have a 100% survival rate."

To promote growth, every two to six months new side branches and shoots are removed.

property. There were many reasons for this resistance. Often staff were outsiders, not from the village, though in some instances this could be an advantage. Some were Christians, and in some cases, villagers didn't want to work with them because of this. There were individuals who were perfectly happy to keep stealing other people's trees. Some simply felt threatened by change. Importantly, my staff and local champions persevered and ultimately won the confidence of communities and influenced a whole generation of farmers.

Early resistance

While I had no doubt that I had found the solution, the discovery of the "underground forest" didn't take Niger by storm. In fact, acceptance of FMNR was slow at first. The paradigm of a good farmer being a "clean" farmer, one who clears trees from his land, was deeply ingrained, and nobody in that culture wanted to be seen to be questioning or going against accepted wisdom. Farmers had been clearing trees for generations and most could not see the wisdom in allowing trees to grow in their valuable fields. Or, even if they did, because they didn't own them, they knew that it was more likely than not that somebody would steal them. So why bother?

From weed to cash crop

This issue of tree ownership needed to be tackled head on. My staff and I presented the case for change to the head of the Maradi District Forestry Department. I reasoned that punishing people for cutting trees down had clearly failed as a deterrent and pleaded the case for greater autonomy for the farmers. He gave a us a verbal commitment that, as an experiment, the application of the rules of tenure would be relaxed so that farmers would have an incentive to nurture and care for the trees on their land. He had nothing to lose. Most of the trees had already disappeared.

Our project staff began telling farmers that they could harvest trees without the fear of being fined. Farmers could pay a small fee for a permit to sell fuel wood at local markets. As they began to see trees on their farms as a cash crop rather than weeds and a liability, farmers were motivated to look after them and protect them from theft. Over time, village and district chiefs established new codes and rules. This greatly facilitated the spread of FMNR. However, the national laws governing tree ownership and user rights would not change for another 20 years! It was a clear case of policy following reality on the ground and not the other way round.

The Forestry head's verbal commitment helped but, sadly, it would take a famine to change the cultural perspective on trees.

CHAPTER EIGHT
May your God help you

The Hausa are a proud people. As the population grew, villages were established further and further from the fertile flood plains onto the sandy plains. The forefathers of the present generation cleared the dryland forest to make way for farming. In those early days, wood was abundant—so much so that fencing around home compounds consisted of a row of upright logs, side-by-side and partially buried in the ground. Such fences were very rare by the time I arrived—most of the wood had been used as fuel wood and replaced with grass mats or millet straw. Farmers had larger farms, and sometimes surplus grain was left unharvested in the fields. People could meet their needs from the natural resources at their disposal supplemented by purchases in the bush markets.

Starting in 1968, rainfall patterns in the West African Sahel underwent sudden and extreme change. Mean August rainfall in our operational area for the period from 1968 to 1997 was around 37% lower than the mean rainfall for the 30-year period from 1931 to 1960. Slightly further to the north of our operations, the reduction in rainfall was 55%. Such a drastic decline in mean rainfall values over several decades is probably unique in the world.[28] Rainfall was also variable year to year and even within seasons. Yet, farmers continued to rely on annual crops under rainfed farming conditions. As a result, they were vulnerable to crop failure and hunger.

The removal of trees intensified the impact of drought. Without the protection of grass or crops and without the shelter of trees there was nothing left to hold the soil in place.

The loss of perennial woody trees and shrubs compounded the problem. Trees play a critical role in reducing temperatures, wind speeds and evaporation. Trees help maintain soil fertility through a number of mechanisms. Some tree species even exhibit hydraulic lift.[29] This is a process through which tap roots draw water from deep in the soil profile and at night time release

During severe drought of 1984 dunes invade an abandoned village. Villagers had left their homes in order to find food and work in towns, cities and neighbouring countries.

it into shallow soil surface layers within reach of the roots of nearby crop plants. Effectively, crop plants are "bio-irrigated" by trees and bushes, protecting them to some degree from drought. Trees also provide habitat, shelter and food for natural predators of pests such as insect-eating birds, spiders, agama lizards and chameleons. The loss of trees resulted in reduced habitat and hence reduced predator populations which enable insect pests to breed rapidly and wreak havoc in crops.

While climate change has complicated the situation, it should be noted that in semi-arid zones, periodic drought is normal, not an anomaly. To this end, reliance on annual, rainfed crops alone is risky, and is in fact, a recipe for disaster. Biodiversity should be maximised to the degree it is practical. Shrubs and trees have deep roots and can ride out variations in rainfall by relying on moisture deep in the soil profile, at least producing something, even in a dry year. As well as buffering extremes in climate, various tree and bush species provide edible leaves, fruits and nuts, fodder, wood for fuel and building, and more. In this way, even when staple crops fail, farmers will have something to fall back on. By including perennial shrubs and trees as well as livestock, farming systems can be made more robust.

Given the almost complete loss of woody vegetation and the variable climate it was inevitable that a crisis would come. *In ka ga kunnen jaki, wutsiya tana baya*—If you see the ears of the donkey, you can be sure the tail will follow. Each rainy season my heart was in my mouth as I looked to the cloudless sky and wondered when it would rain, if the crops would survive

Even the granaries had a story to tell. In 1984, most grain silos were empty; a sure sign of a poor harvest the previous year and hunger. In desperation, some farmers sold their silos for a pittance in order to buy food.

and if people would have food. In 1984, my worst fears were realised. The rains failed completely. By the end of the "rainy season" millet plants, which should have been two metres tall and bearing spear-like grain heads, were no more than a mass of crumpled, brown leaves. Grain production was minimal. People were already living on the edge of hunger as the years preceding 1984 were also marginal, leaving many with no grain reserves. There was no buffer. I learnt that when it doesn't rain, no food is produced, and when there is no food, real people, people I knew and loved, not just unknown people from the other side of the world on a TV screen, go hungry, and some starve.

In 1984, as hunger again stalked the land of Niger and, indeed, all of the Sahel, Richard St Barbe Baker's prophetic warning on the dire consequences of destroying forests seemed to be fulfilled before my eyes, "Then the age-old phantoms appear stealthily, one after another—flood, drought, fire, famine, pestilence." St Barbe Baker's 40,000-kilometre Sahara challenge expedition in 1952 was partly in response to a shocking story from upper Dahomey, in modern-day Benin. A chief had forbidden his subjects from having children because of the encroachment of the Sahara and looming famine. St Barbe Baker, a pacifist, wanted to mount a multinational Sahara reforestation campaign. On the outskirts of Tamanrasset, southern Algeria, he saw woodchips on the ground. They were a result of felling trees with axes, and tantalising evidence of recent remnant forests in what was effectively the heart of the Sahara,

evidence that the forest could be restored. He passed through Zinder, 235 kilometres to the east of Maradi before heading south to Nigeria and then east, all the way to Kenya.

Famine

As the months of hunger ground on, horror stories began to emerge. In order to survive, people ate whatever they could—wheat bran, meat of animals that had perished, bitter and even toxic leaves and fruits such as hanza, (*Boscia senegalensis*), which had to be boiled several times, changing the water between boiling, to remove toxic alkaloids.[30] Tragically, occasionally you would hear of people dying who did not know how to treat the fruit.

People, particularly men, were leaving their villages in increasing numbers, looking for food and for work in order to be able to support their families. Generally, women and children stayed at home, coping as best they could in the hope that their men would send food or money. When all options were exhausted, they too made the long trek to major centres. Pitiful columns of women carrying their sleeping mat and meagre possessions formed; they walked 20, 50, even in excess of 100 kilometres in order to survive. Some were pregnant. Many carried babies and had toddlers in tow.

I wondered what was going through their minds. Death and suffering were regular companions in Niger even without total crop failure. They hoped to find work and food, to survive and to scrape enough together to be able to return once the rains started. Then they could cultivate and sow their fields in hope of a better season the following year. This was imperative otherwise their desperation would extend even longer.

Because they had run out of food, these women collected the poisonous fruit of the *Boscia senegelensis* tree which had to be soaked and boiled for two days until it was edible.

With so much excess labour in the cities, already low wages plummeted even further. Famished women contracted for cents to shell peanuts were punished if they were caught eating any. Women sometimes had the Faustian choice of selling their bodies or letting their children starve.

In a remote village to the northeast of Maradi, the men left first, hoping to find and send food home. Gradually most of the women and children also left. Safiya had five children. She did the best she could, boiling up weeds, bran, even the leaves of neem trees, which had to be boiled with several changes of water to remove the bitterness. She hoped against hope that her husband would send food or come and rescue them. But he never came. Finally, Safiya ran out of food altogether and she was the only one left in the village. Making the 100-kilometre trek on foot with five young children would be impossible, so she locked three in her hut, and walked to the nearest big centre with the remaining two. When she was finally able to get help to return, it was too late. The children had died in the hut.

I heard stories of suicide, otherwise rare in Niger. When a proud Fulani herder lost all his livestock, he threw himself into a 60-metre deep well. One day I came across a blind man being led by a boy holding onto his walking stick. He told me that he had walked 70 kilometres from his village in the hope of finding food. This was the fourth failed season in a row and things were getting desperate. People were leaving; his neighbour had committed suicide. His plight moved me. I loaded grain for him and his neighbour's family onto the pickup and drove him back home. The scene in his village was surreal. Sun shining through the dusty atmosphere gave everything—earth, huts, vegetation and people—a Martian red glow. There was no sound of pounding grain, no sign of life. There were no normally ubiquitous chickens—an edible and cashable resource—in sight. Men were reploughing their dry, dusty fields. Were they hoping that the heavens would see their plight and that God would have mercy and open the rain vaults of heaven? Or perhaps they engaged in this futile exercise to escape their children's cries of hunger.

One day I visited a good friend, Dan Bakwai. We drove to his millet field, which in a good year would have been green. The field was bare. There was nothing but windblown sand. The ground was bright and hot. Dan Bakwai walked on ahead of me. His shoulders were slouched and he muttered to himself. "How will I feed my four wives and 15 children now?" I feared for his sanity. This was the third year in a row he had planted a crop and harvested nothing.

Mohammad and Gwamma and their three children lived in a village to the North of Maradi. As the family grew and grain yields declined, Mohammad found it harder and harder to make ends meet. It is the husband's responsibility to provide a new three-piece outfit for his wife each

"How will I feed my four wives and 15 children?" This was the third year in a row my good friend, Dan Bakwai, had planted a crop and harvested nothing. As he wandered off, muttering to himself, I feared for his sanity.

year. For a wife to wear worn clothing was a thing of shame and a source of tension in the home. The husband is also responsible for providing grain. Apart from the pain of seeing his wife and children go hungry, there is considerable social stigma associated with not being able to provide for your family. It is the wife's responsibility to draw water, collect fuel wood, care for the children and small livestock and to provide the condiments necessary to make the otherwise plain millet porridge more tasty and palatable. As the environment and climate deteriorated, it became harder and harder for Mohammad and Gwamma to meet each other's expectations. The family sometimes reduced meals, such as they were, to twice and sometimes only once per day. Muhammad spent more and more time away from home looking for food and work, in order to send something back home. With Mohammad away, all of the farm work, particularly the back-breaking job of weeding the fields, fell on Gwamma's shoulders. Paid work is hard to find in Niger in the best of times. During times of scarcity, it is almost impossible and those with money take advantage of the desperate. When she did find work, Gwamma was paid with only a bowl of millet bran (traditionally fed to the animals) for a whole day of hulling peanuts.

Even in "normal" times, a vast army of Nigerien men migrate to Nigeria every year as soon as the harvest is in, in the hopes of finding work. Some stay away for the length of the dry season—six months, eight months. Some are never seen again. This famine amplified what was already commonplace.

As for all women left behind, the loneliness and burden of managing the house and farm alone took a heavy toll on Gwamma's health. Long hours of hard physical work, poor nutrition, the stress of not being able to meet the demands of young children and of not hearing news of Mohammad for months on end contributed to Gwamma needing medical attention— attention that required money that she did not have. No wonder she resorted to traditional healers— they were trusted, accessible and affordable. Unfortunately, sometimes their cures did more harm than good. The bark concoction that Gwamma took irreversibly damaged her liver. She and her children often suffered from severe stomach cramps from the bitter leaves they were eating. Bran, which is normally fed to animals caused excruciating constipation.

The economic strength of rural Hausa women lies in the small livestock which they keep. When Gwamma sold her last goat to pay the medicine man, the humiliation of being one of the few women in the village to not have any animals was more unbearable to her than the poverty itself. Although of little consolation, land degradation, being the great equalizer that it is, eventually forced most of her neighbours into the same desperate situation.

Meanwhile in Nigeria, the great influx of foreign workers provided an easy target for ambitious politicians wanting a scapegoat. The ailing economy was not their fault but that of foreigners who were stealing local jobs. Mohammad was forced to hastily join throngs of other evicted Nigerien men leaving the country by whatever means possible. Many exited Nigeria ignobly, standing crushed together like cattle with their meagre possessions in semi-trailers. The exodus involved passing through many roadblocks and having to pay corrupt officials money from his scant earnings in order to be allowed to pass through—money that Mohammad needed for his family.

We were able to help many women and malnourished children but we could not help everyone.

The best they could do was to scrape enough money together to plant their field with the first rains, then bundle their meagre possessions up and move to the rapidly growing "temporary" settlement on the edge of Maradi, with huts constructed of cardboard, thin wooden poles and old cloth. The cost of going back and forward to tend the crop meant that weeding would not happen in a timely manner and the weakened crop yielded a poor harvest.

In order to feed his family, Mohammad had taken a loan—one bag of grain now, in return for two bags as payment at harvest time. When Mohammad defaulted on the payment because of the poor harvest, he forfeited his only remaining asset—his farm—and along with hundreds of neighbours transformed their temporary urban housing arrangement into a permanent one.

> LETTER HOME. 16 NOVEMBER 1984. The situation is very serious at the moment. People in the villages are living on wheat bran and tree leaves—if they can get them. Many animals have died or have been killed early and eaten. There is no grass. In some areas the crops completely failed. Only one or two rains were received. At the Farm School, crops headed out but instead of 1,000 kilograms to the hectare students harvested as little as 300 kilograms. Every day people are coming down from the north looking for food or work in Maradi. Others pass on to Nigeria on foot with all the possessions they have left. They don't know what waits for them. Many people have sold all they own to buy food—their farm implements, clothes, cooking pots, livestock. The problems will be worse later into the dry season.

Surrounded by hunger

Seeing this stress and suffering impacted me deeply and changed me forever. I resolved not only to address these immediate issues, but to tackle the long-term root causes.

When I was a boy, reading about hunger or viewing it on TV left an intense emotional burden on me. But since I didn't have a personal connection, it all seemed remote. By 1984, I had been living in Niger for nearly three years. I had made good friends. I had received their hospitality, slept in their huts, talked late into the night around a kerosene lantern, eaten their food. I knew their children, their joys and their sorrows. When they came to Maradi, they stayed at our home and we reciprocated their hospitality. I had come to love and respect the people. Now my friends were hungry. This was the defining moment in my life. What was I to do? I didn't have any experience in famine relief. I could crumble and say there was nothing I could do about it. Or I could at least attempt to obtain help and provide relief, as daunting as that was. In the end

A common sight in 1984. First the men left their villages in search of work in order to send food or money home. Then, when the food supply ran out completely, the women and children followed with what few possessions they could carry.

I could not stand by and do nothing. My friends, and many thousands of others, were suffering all around me.

I did not receive much encouragement from some quarters. One NGO leader didn't want to get involved because it was time-consuming and would disrupt his organisation's normal programming. What was the use of development programs that didn't address the serious and immediate needs of the people we are trying to help? One government official in the Ministry of Planning said, "May your God help you," implying that he wouldn't. Even some of my colleagues outright opposed me and believed it wasn't my role or the role of the mission to distribute food. However, local, regional and international SIM mission leadership were very supportive. John Ockers understood my situation. He had pioneered many innovations and had been heavily criticised for them. "The only people who are not criticised are those who do nothing. Only the go-getters and world changers face opposition."

No one taught me more about Hausa culture, language and character than my Nigerien counterpart from the Evangelical Church of Niger (EERN), Pastor Cherif Yacouba. The Maradi Integrated Development Project was managed by SIM and EERN so Cherif and I were jointly responsible for its operation and outcomes. Pastor Cherif came from a chiefly family. While he wasn't a chief himself, he was self-assured and confident. His manner and the way he conducted himself conveyed authority and commanded respect and his standing in society opened doors and made it possible for me to do many of the things I did. Authorities listened to him and often gave him what he wanted. In the eyes of village people, he was a big man. When we needed to get things done, Cherif was direct and didn't shy away from conflict and difficulty.

Pastor Cherif and I decided to seek an audience with the *Préfet* (governor) of Maradi Region to ask him for grain and permission to distribute food. Pastor Cherif had a proverb for every occasion. "*Zuwa da kai, ya ba kura baban wuya.*" (By going himself, the hyena got its thick neck.)

The *Préfet* refused to give us permission to distribute food and denied us access to the government grain reserves which had been specifically established for emergencies like this. This was a major obstacle. Without his permission we could do nothing. Strangely, even the merchants in the market didn't seem to have grain they could sell to us. In any case, we had no funds. I began to pray that the *Préfet* would have a change of heart or be transferred.

Each day, more and more people congregated around our house seeking help. I wrote in my daily journal:

> Dear Lord God, here I am a prisoner in my own house, surrounded by a sea of hungry people. What should my attitude be? How can I show compassion?
> There are so many! They wait for me day and night; they have nowhere else to go.

They began sleeping on the ground overnight in order to catch me early in the morning before I left home. The mission had distributed grain during the 1970s food crisis and people hoped we would do it again. It was emotionally draining to be surrounded day and night by desperate people with starving children, and not to have the means to help them. Each morning I would have to push my way through the crowd to get to my vehicle. One morning, Ben, who was three years old, followed me out the door. As I slowly progressed through the crowd, I was surprised to hear, behind me, Ben shout with a confident, clear voice. "*Ba ku ji ba? Ya ce ba ya da abinci. Bar shi, shi wuce!*"—"Didn't you hear him? He said he didn't have any food. Let him pass!" I was able to pass through the temporarily stunned crowd. Ben seemed to understand my plight. I couldn't help them by just talking. At least I was trying to do something and that somehow made the situation more bearable for me. It was perhaps more draining on my colleagues who were also constantly surrounded by needy people.

The obstacles seemed insurmountable. I was having a crisis of faith. As the pressure mounted day after day and I was unable to help, I wondered, did God really care? Did he really answer prayer? Could he really use very ordinary people to make a difference in impossible situations like this? I was also having a crisis of confidence. What did it mean to be a Christian living in the place where people were suffering? If I didn't help them, then I might as well go home immediately as my words about a loving God would have been empty. Even so, I wondered how I could find practical solutions for these myriad problems. Seeming clichés, such as "God loves you," "God answers prayer," are easy to recite in quiet meditation and with other Christians, but it's only in the crucible of life that they, and we, are tested.

Again, God answered my desperate prayers for help. One morning I was trying to eat my breakfast. As tears were welling up and a lump was forming in my throat, my eyes fell on the following verse in the open Bible, "Thus says the Lord to you, 'Fear not, and be not dismayed at this great multitude; for the battle is not yours but God's.'"[31] Immediately a great weight was lifted from me. I didn't know what, how, or when, but I felt an assurance that God was going to act decisively.

CHAPTER NINE

The lizard also drinks from the chicken's water bowl

Within two weeks, the Maradi *Préfet* died. The mayor of Maradi, an old friend of the mission, became interim *Préfet*. He immediately gave our EERN–SIM team authorisation to buy grain from the government reserve and distribute food. One colleague found the change of fortune amusing and said "Maybe the Mayor was scared that you would pray for him too!"

Mysteriously, grain suddenly became available on the open market. Over the following months we received half a million US dollars mostly from individuals and churches donated through SIM offices in developed countries. We were able to distribute 1,800 tonnes of grain to over 70,000 people. God does care. God does answer prayer. God does use ordinary people in impossible situations.

The government of Niger wisely decreed that no food handouts were to be given to able-bodied people unless they worked. FMNR became a beneficiary of this policy. If we were going to distribute food we were obliged to also give people work to do. Cherif and I looked at each other. "This is our chance!" Our team initiated a food-for-work program. One of the "work" requirements was that each household receiving food aid must regenerate and maintain at least 40 trees per hectare on their own land. Most farmers were very reluctant about this at first, but they agreed to do it. The program gave us an opportunity to introduce our new strategy to a critical mass of people for an extended period so that they could experience the benefits for themselves. As with our earlier experiments, FMNR worked extremely well. For the first time, reforestation was occurring rapidly, at low cost and at scale. Many were surprised that their crops grew better amongst the trees, even from the first 6 to 12 months of regrowth.

> The food-for-work program saved lives and exposed people to the benefits of FMNR.

All benefited from having a small amount of fuel wood and fodder as a product of pruning and thinning the regrowth. Under the protective umbrella of the food-for-work program, and the support of village and district chiefs and the forestry service, the stigma of being different was suspended and theft of trees was kept to a minimum.

Grain distribution to the villages

> DIARY ENTRY. 18 JANUARY 1985. Time is really flying. We haven't kept still. No sooner do we do one food distribution, than there is another. Thank you that the food is coming in Lord. What is relief work like? It's like trying to stay on top of a fast-moving wave; momentarily one gets to stay on top, but mostly, I'm either underneath it, gasping for air, or paddling flat out trying to catch up.

> DIARY ENTRY. 22 JANUARY 1985. Dear God, please help. I have that sinking feeling in the pit of my stomach. The whole thing is too much. I am not adequate for this task. Please help me to be organised. Help me to balance the books and the in and out tally of grain. Have FAITH, not fear. Faith overcomes fear. When I got back home there were more people at the door; for five days we had no food to give the Fulanis. Always people asking me for food and work. More work to do than time to do it in. Big problems at Soura. Too many hungry people. Help us Lord. May I trust you for today—that all that needs to be done will be done. May I be faithful. May I not neglect my family so much.

Despite the sense of urgency, mounting such an undertaking involved lots of waiting: in government offices for permissions, and for funds and for grain to arrive. All this took enormous effort. There were many tasks: travelling to each village, which numbered over 100 by the end of the relief period, making distribution lists, looking for, purchasing, transporting, storing, guarding and distributing grain. As this was before computers were widely available, I manually created large spreadsheets on paper which covered our dining table. They contained names of villages, heads of households, numbers of family members, grain rations per month and work details. We travelled hundreds of kilometres each week, often on corrugated roads under hot and dusty conditions. Even Pastor Cherif found it exhausting. *"Hanyoyin nan suna zazagen cikoken tsofofin!"*—These roads are rattling the stomachs of the old people!

> DIARY ENTRY. 23 FEBRUARY 1985. Spent all of yesterday at the government office in Dakoro, waiting for the signature of the *Sous-préfet* (sub-governor).

Providing famine relief was totally new territory and I did not have a manual. I had so much to learn about logistics, organisation, crowd control, record keeping and human nature. My first purchase of grain—40 100-kilogram sacks, or four tonnes of grain in total—seemed an enormous accomplishment. It took a whole week to distribute it as we transported it to each of the dozen villages on our list. However, my satisfaction vanished when a recipient immediately asked me when the next delivery would arrive! How naïve I was. Four tonnes were a drop in the ocean. I was not finishing a job; it was just beginning!

Our team reorganised. Delivering food to each village clearly wasn't going to work, so we selected four distribution centres and stored enough grain for a monthly distribution. The day before each distribution was to happen, we sent messengers to inform the villagers to come and receive their allotment. This system was efficient and effective.

Distribution days were always very taxing. The main task was to measure out the correct amount of grain for each family registered and to do this as quickly as possible. And all this in the heat, surrounded by people pressing in, all eager to get their share, or more. Regular exposure to grain dust resulted in an itchy rash on contact and fits of sneezing. At each distribution point, unregistered people turned up in the hope of being fed. This complicated things further. Even though we were hard pressed to provide enough grain for those on the list and ensure its timely delivery, we couldn't just ignore people. They had to be treated with dignity. Following one particularly arduous distribution I returned home dehydrated and with a splitting headache. I lay down on the living room floor and asked Liz for something to drink. My headache only started to subside after I drank two litres of water, a Coke, a Sprite and two cups of tea. But I felt satisfied. The food was getting through to those who desperately needed it.

The responsibility weighed heavily on me. I could never be absolutely sure that I was helping the most needy. I was angered when I learned that people lied about their assets in

In 1984, we distributed 1,800 tonnes of grain to over 70,000 people.

order to receive grain, and I was frustrated when two different women turned up on different days with the same ten children to be registered. And there were times when I was completely gutted to learn that I had denied help to people living in extreme circumstances. I walked a fine line between the arrogance that can come from having such power and humility from knowing people's welfare lay in my hands and that I would make mistakes. One day somebody was being particularly exasperating, and I got cross with him. Immediately I felt convicted: who did I think I was? Would that person bother me if they weren't hungry? If there was ever a time for crying out for courage, wisdom and humility, this was it.

    The utmost vigilance was always required. Theft could happen anywhere and at any time. I stored grain in Maradi on SIM mission compounds and in rented storerooms. I purchased large tarpaulins so that I could store grain in the villages where distributions would take place. I paid guards and counted the bags going in and coming out. Cheating could happen at unloading. I caught casual day labourers who were unloading a truck filling their shirts with grain that had spilled from broken bags. They were throwing their bulging shirts over the front of the truck to their friends, who emptied them and threw them back to refill. Even the occasional Bible school student or evangelist would succumb to temptation. I can't judge them. Only someone who lives in poverty can understand the enormous pressures and cultural expectations extended family members can exert. However, I was responsible for doing everything possible to stop it happening. Often theft would occur at distributions, so I regularly changed the roster so that workers had no time to work out a plan beforehand and I did not announce the date of distribution until the last possible moment. As individuals worked out ways around the system, we would have to come up with new strategies for the next round.

The biggest disappointment came from one of my own staff members. He was responsible for making the distribution lists. When he was approached by people from villages not within our working area, he saw an opportunity to make money. Registration was free, but he demanded payment in exchange for registering them. His greed was his eventual undoing. Not content to collect money illegally, on distribution day he also withheld the grain to sell later! This happened on three successive distributions to one of his victims. However, the third distribution fell the day before a religious feast day, and the victim had vowed he would not come away empty-handed. When my staff and I emerged from lunch break, the victim was indignantly standing his ground—literally standing on his bag of grain—and refusing to move until he received his due. The scuffle between him and the guard exposed the fraud and we were able to sack those involved.

Nothing ever went completely to plan. One morning while we were setting up for a distribution in a rural location and hundreds of people were arriving, a distraught father brought his daughter to me to take to hospital. She had fallen into a large pot of boiling water. I did take her, but to my shame, only after everything was set up and I was satisfied nothing could possibly go wrong. Hours later the father returned to tell me that his daughter had died.

This was a particularly difficult time for us as a family. While I was out, Liz had the pressure of having starving people gathering just outside our gate all day. Ben and Melissa were distressed that I was often away and rarely arrived home before they went to bed. Liz had to borrow a car from friends to go shopping because when people recognised our pickup truck, they would hop on the back and beg to be taken to *Malam* (teacher) Tony.

In need of a break, in December 1984 we planned to travel east to spend Christmas with fellow Australians, Phil and Carol Short and their children. But first I had to organise the purchase of grain. Believing that the paramount chief had grain for sale stored in Maradi, we thought it was just a matter of us hiring a local truck to pick it up. I never expected him to lie to us blatantly, so when he insisted we pay in full in advance we foolishly complied. My heart sank when he proceeded to order his agents to cross the "closed" border and buy the grain in Nigeria.

The famine of 1984–1985 was locally named Yar Buhari (Daughter of Buhari) because President Buhari of Nigeria had banned all grain exports in order to maintain stocks for his own people. This contributed to soaring grain prices in Niger. Actually, nearly all the grain reaching the markets in South-Central Niger came across the Nigerian border illegally and usually at night when bribed customs officials could turn a blind eye. Adding injury to insult, weeks after our purchase, but before delivery was made, the chief maintained that the grain sacks weren't included in the price. He insisted on us paying for new sacks—something which is never done in Niger. When the grain finally arrived, it was contained in torn and patched bags! I learnt my lesson the hard way on that occasion and was very wary of him from then on.

The grain for distribution arrived on Christmas Eve. We could not travel, knowing that we had food while people were starving, so Liz, Ben and Melissa did what they could to celebrate while I was gone from before dawn to late in the evening. We had no way of informing the Shorts as there were no telephones. That day was windy, and the dust made the sky an eerie red colour. On the way to the villages several 100-kilogram sacks fell off the pickup and spilt on the laterite road. Not having any shovels, my workers and I suffered from gravel rash as we scraped up the grain from the road with our bare hands.

Grain distribution to the Fulani camp

Throughout the entire period of famine relief, Pastor Cherif and I were constant companions. We rose early in the morning and worked late into the evening. Whether addressing chiefs or government officials, he was a fearless advocate for the poor.

One evening, Liz and I were having a meal with senior missionaries Gordon and Lena Bishop at their home in Soura, just three kilometres from the Farm School. There was a commotion outside, and Gordon and I went out to investigate. A Wodaabe Fulani family had arrived on donkeys with all their possessions. Gordon welcomed them and asked, "Where are you going?" The response was, "We have arrived!" A trickle became a flood and within a few months there were 2500 needy nomads camped there with no intention of moving on.

Tragically, Gordon was crushed to death in a well accident in October 1984.

Pastor Cherif and I were able to secure food from the World Food Program and from the market, using funds sourced by SIM. We organised for shallow wells to be dug, purchased seed and hired staff to teach gardening skills to nomadic herders who had never held a hoe in their lives. When he tasted boiled potatoes for the first time, one Fulani elder commented "Mmm, this is good. It's ... it's almost as good as milk!" Milk and milk products are, of course, the staple of Fulani cuisine.

It was chaotic. Nomads proved to be extraordinarily difficult to organise. Somewhere along the line we realised there were discrepancies in our lists of names, so Pastor Cherif insisted on checking that everyone registered was physically present in the camp. We worked until past midnight as each family group paraded through our lean-to and we checked off the names on the list by the light of kerosene lamps, perspiring and bombarded by kamikaze insects attracted to the light. This exercise indeed routed out a number of false claims:

Pastor Cherif to husband: "Where is your wife?"

Husband: "She passed away."

Pastor Cherif: "Oh, I'm so sorry. When did this happen?"

Husband: "Two years ago."

Pastor Cherif: "But you are collecting grain for her! So, what do you do, throw her grain ration to her up in heaven?"

A measles epidemic broke out in the camp and it took weeks to get the local clinic to vaccinate, by which time over 62 children died. In the meantime, someone with an axe to grind used this as a pretext to undermine the work and reported to the *Préfet* that we had detained Fulanis at Soura and were starving them to death. The next day, the gendarmes, the *Préfet's* Secretary General and Secretary to the Mayor arrived, ready to arrest us. When they saw what we were accomplishing with the meagre resources available they apologised, praised our work and provided milk, vitamins, a tent and blankets. What was intended for evil turned out for good. We encountered a similar situation at one of our distribution centres for the village of Batafadua. The *Sous-préfet* for the district received a false report that "the Christians held two distributions: a general one for everyone, and then later a special one for those who had converted to Christianity. The *Sous-préfet* turned up in person on a distribution day to verify the charge and ended up praising us. Following that incident, we were provided with an official witness to accompany us at each distribution.

The same accuser from Soura later went to Pastor Cherif's home with a knife hidden in his boot and threatened him. When some people can't work the system for their own personal gain, they become hell-bent on destroying it for everybody else.

Feeding centres

> LETTER HOME. 31 MARCH 1985. We are still trying to set up a feeding centre for women and children. Liz is doing what she can for 20 women with malnourished children until the centre is ready. Each day more women arrive but we don't have the resources to help any more. We've started padlocking our gates. Each day we have up to 150 women and children in the yard outside our fence. It's very stressful and disheartening not being able to help them yet. We hate turning back the ones we can't help when we know they need it, especially the ones with bony children who blankly stare at you through their beautiful brown eyes. I have confidence that we will help them, but things take time and we already have a busy schedule helping out in the villages.

During later periods of famine relief in Niger, I set up a number of temporary feeding centres for malnourished children in the villages and adjacent to the property we lived on in the city of Maradi. At age 15 Melissa spent part of her school holiday helping in one of these centres.

> I met lots of mothers and babies during my time at the centre, but one particular baby left a great impression on me. I was stunned when I was told how old she was because she was so very small, far too small, but so very beautiful. This baby was severely malnourished. I haven't forgotten the big earrings she was wearing. They seemed huge and solid compared to the rest of her tiny, frail body. "What a beautiful girl," I thought.

Thanks to the food-for-work program, farmers in 100 villages regenerated 500,000 trees.

> "But she won't make it. I know she won't." I willed her to live. I willed for the help she received to work. A few days later I returned and asked after her and she had died. The picture of that little girl has never left my mind. Ben remembers her too.

A watershed year

The 1984–1985 famine was a watershed event and became a turning point in my life. The pre-famine Tony and post-famine Tony were two different people. Seeing God answer prayer in such a powerful and unambiguous way transformed me. Nothing in my life till then had galvanised my focus and energies in the way this crisis did. While I worked hard before 1984, I now felt energised with renewed vision and purpose. On the one hand, I saw the unnecessary suffering of people dependent on a degraded landscape that could no longer meet their needs. On the other hand, I saw the enormous untapped potential of FMNR to restore landscapes, to make them productive again, preventing famine from happening in the first place. Like the proverbial hungry chicken, despite appearances, people in Niger were sitting on a veritable granary. If farmers worked with nature instead of destroying it, the land was capable of feeding the current population and more.

While urgently getting food to hungry people was the top priority, in the long run it would be irresponsible not to also address root causes of hunger: deforestation and land degradation. "Never let a good crisis go to waste!" said Winston Churchill. Hunger, in this case, was a symptom of deforestation. I was determined to bring long-term good out of a bad situation, and the effort paid off. Thanks to the food-for-work program, 70,000 people in 100 villages regenerated approximately 500,000 trees across over 12,000 hectares. This wide-scale exposure to practising FMNR proved to be a turning point. The fact that good rains in 1985 produced a bumper crop, despite there being trees in the fields, was not lost on many farmers. The doubling of average crop yields that year supported the narrative that trees and crops are compatible.

At last, it seemed that we had a viable means of restoring Niger's degraded landscapes. However, things don't always go to plan. As soon as farmers began harvesting their own grain in September, we discontinued the food aid but continued teaching on the importance of trees and encouraged people to practise FMNR. But old habits die hard. It is hard for poor people to wait for a tree to grow when they can make quick money today selling firewood. Many harvested their trees in order to avoid having them stolen. And others still didn't believe that trees and crops were compatible. When the food handouts ceased, over 75% of approximately half a million trees cultivated in 1984 were chopped down! People felt they could now get on with their lives and go back to the way things were without my interference.

Even so, the harsh reality of famine sharpened the thinking of enough people about the importance of trees. Perhaps for the first time they began to associate the increasing frequency and severity of droughts with the loss of trees? The farmers who didn't immediately clear their newly regenerated trees still comprised a critical mass that would go on to sway public opinion. More and more farmers started protecting trees and word began to spread from farmer to farmer until it became a standard practice. The seed of a movement had germinated.

CHAPTER TEN
Year of the Tankarki

The years from 1987 to 1999 were busy but fulfilling. I had no doubt about the power of FMNR to transform lives and landscapes. I found time away from the project—for French language study, a stint as Eastern Niger interim director for SIM and periods of "home assignment" every four years—frustrating, but I never lost the vision and the passion.

Recurring hunger

During this time, farmers faced four further significant food-shortage years. We did not let that defeat us but turned tragedy into good by implementing food-for-work programs designed to feed people while addressing a major root cause of food insecurity: deforestation and consequent land degradation. Significantly, as adoption of FMNR spread and the trees grew, farmers became more resilient. With each successive food crisis, the amount of food aid needed declined. A food crisis in 1988 revealed nature's fury and capacity for the destruction of our life support system. It should stand as a warning to those who intentionally and shamelessly continue to contribute to climate change through their policies and investments. Sadly, there are and will be severe consequences for us all, but more so for the poor.

Halidou Gangara proudly displays his FMNR trees growing in a millet field. Firewood can be seen in the background. Direct benefits from growing trees reinforce farmers' willingness to adopt FMNR.

In 1988 a benchmark was set for retribution wrought by an ecosystem which has been violated. The rainy season started late. When rains did come, they were weak, but farmers planted in the hope of follow-up rains. Severe winds buried and sand blasted emerging crops, destroying them. Waves of grasshopper nymphs marched across the fields like miniature lawnmowers devouring every blade

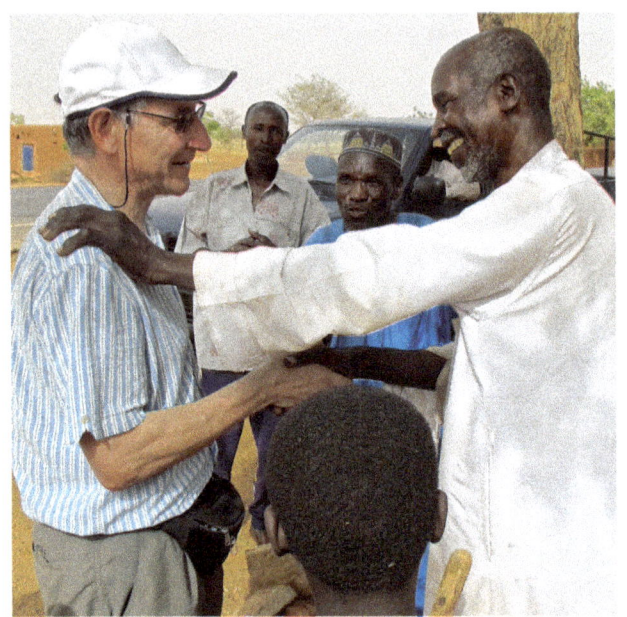

The person most changed by our time in Africa was me.

of millet, the main crop and staple food. Crows, and in some districts, desert rats, hopped directly to each drill hole and ate the buried seed. Farmers replanted precious seed, which was also their scarce food supply four, five, six times in some districts. The millet that survived this onslaught was attacked by stem borers, small caterpillars which burrow into the stem of the millet plant, greatly weakening it and reducing the yield. The plants that survived that attack and went on to form grain heads were attacked by head borers, caterpillars that specialise in eating grain. Not content to eat their fill and leave, in the process of burrowing beneath the rows of millet grain, they waste as much grain as they consume. In any case, any vegetation that had survived these setbacks eventually desiccated as the rainy season ended before the grain could fully ripen. With no habitat for natural predators such as insect-eating birds, lizards and spiders, pest numbers exploded, destroying crops. Farmers, who had few options of livelihood other than what they could grow on their land, became destitute.

Village heads called a meeting to discuss the looming famine and possible solutions. They were painfully aware of their situation. There was no new ground to clear and cultivate. To the north there was desert. The countries to the west were also suffering from drought. To the east there was political instability and to the south they were not welcome. At home there was hunger.

When I was asked to speak, I told a popular Hausa fable.

"Once upon a time, Zakara (the rooster) invited Tankarki (the wild bustard bird) to spend the day with him. They had such a wonderful time that at sunset, Zakara asked Tankarki to stay the night. Tankarki was unsure about this. Looking into the chicken coop he noticed there was only one way in and no other exits. Tankarki asked Zakara, 'How do you escape if Mutum (the man) puts his hand in here?' Zakara looked down. 'There is no escape.'"

Everyone knew how the story ends.

"Unimpressed, Tankarki flew up into the trees crying, *'Hauka!'* (That's madness!)"

We could see no other way out of the hunger trap. The leaders assured me that they wanted 1987–1988 to be the year of the Tankarki, and not the year of the Zakara!

An older missionary once recounted how his family fled Bolshevik Russia. Escaping by night, his grandmother would encourage the family with the words, "When the night is darkest, the stars shine brightest." Periods of famine with the suffering they entail have to be amongst humanity's darkest nights. However, if not "wasted," famine can also bring out the best in people. It can make them shine brightly as they bring about heroic changes that would not have been possible in normal times.

Indeed, from this point onwards we began to witness the emergence of an FMNR movement; slowly, imperceptibly at first, an unstoppable people-led movement changed the way Nigeriens did agriculture.

How many notches do you have on your belt?

What does it mean to be a missionary undertaking development work? Being a missionary is a foreign concept in our largely secular society. Many argue that people have their own religion and should be left alone, as if adults in developing countries aren't capable of making free and informed decisions for themselves.

Sometimes I am asked how many converts I had made, or more bluntly "how many notches do you have in your belt?" It's not like that. Jesus never forced himself on me, and neither do I force him on anybody. I never used the assistance I gave as bait, or to coerce people to convert. That is unethical, un-Christian and counterproductive. If God is who he says he is, he can draw people to himself. My job is to be faithful and obedient, and I leave the rest up to him. If anything, the person most changed by our time in Africa was me.

From time to time in Niger, people I knew did become Christians, like my friends Abou and Bature. From their stories, I'm hard-pressed to know who is responsible for their conversion. Was it a missionary, a Nigerien Christian, the men themselves, God, or a combination?

Abou was devoutly Muslim. His father had built the village worship centre in his home compound. I wanted to employ Abou in our project as an extension agent, but his friends warned him, "Don't work for the Christians; they'll put something in your food, and you will follow them." He was very hesitant but eventually agreed. We always started our team meetings with a brief devotion and prayer but, not wanting to embarrass Abou in any way, I assured him that he was free to attend devotions, or just come for the business part of the meeting.

A member of my staff, Moussa, befriended Abou and mentored him in many areas of life— good farming practices, money management, family life. Religion is not a taboo subject in Niger and during their many discussions Abou asked questions about Moussa's faith. One time while visiting Moussa's home, the call to prayer sounded. Moussa instructed his wife to sweep a spot in the shade and give Abou a mat to pray on. This kindness touched him deeply. When his father had cataracts, I arranged for an operation. The two of them had to stay at our home overnight, so I did what any Hausa host would do for me: I provided food, water and a mat to sleep on. That his boss, an expatriate from another country, would do this for his father also made a big impression on him. On his own initiative, Abou listened to Christian messages on tapes and the radio, and he read Christian tracts.

Abou wanted to know the truth. He was taught that you could only get to heaven through good deeds, but Christianity taught that salvation was by grace through faith in Jesus Christ. He prayed that God would show him the true path. That night, he had a dream. In the dream, there was a man standing on a rooftop calling him, "I am the way, the truth and the life. Follow me." He went to the house to see who it was but as he reached the wall, he woke up without knowing who was speaking. Upset, the next day he prayed again for God to reveal himself. That night he had the same dream and woke again at the same point in the dream. Bitterly disappointed but more desperate than ever to have an answer, on the third day he prayed harder still and even fasted. That night he had the same dream again, but this time Jesus revealed himself. From that point, Abou became a follower of Jesus, despite the opposition and ridicule that he would inevitably experience.

Through his changed life and genuine sharing, Abou went on to influence others. Bature, by his own admission, was a drunkard, an unfaithful husband, wife basher and hopeless gambler. He was always the life of the party but inwardly he was utterly miserable. He saw the change in Abou's life and asked him if his conversion was genuine or whether he had converted just to keep his job. Abou replied, "It is real, and you can have the same peace and the same change of life, but you have to seek God yourself. It's not something somebody else can do for you." Desperately unhappy, Bature prayed, and he also had a dream. In his dream he saw a man in a pit being tormented by snakes. As the man scrambled up the side of the pit to escape, Bature realised that he was the man in the pit. He repented of his sins and believed. If you met Bature today, you would never imagine the life that he once lived.

Genuine faith must come from the heart. I don't hide the fact that I am a Christian. People—especially those whom I help—often ask me why I do what I do. I tell them my primary motivation is my love of Jesus and gratitude for what he has done for me, dying on the cross for my sins. The work of conversion is the work of God in the life of individuals.

A period of expansion

During this period, we took opportunities to promote FMNR locally and nationally. Exchange visits and training days were organised for various NGOs, government foresters, Peace Corps Volunteers, as well as farmer and civil society groups from across Niger. We encouraged them to come and spend time in our villages and learn from our FMNR practitioners. We also sent our project staff and farmers to numerous locations across Niger to provide training. Our farmers may not have been literate, but in my mind they had PhDs in FMNR. Farmers are much more willing to learn from each other than from outsiders. The story of the organic spread of FMNR across Niger is a story of farmers teaching farmers. Taking people out of their familiar village settings and exposing them to a community where FMNR was the norm liberated some to be bold and introduce change at home.

Changing mindscapes to restore landscapes

Hardship makes for an interesting study in human nature. It can bring the best and the worst out of people and lay bare people's true character. It can also pry open the cultural gate that has locked people into established beliefs and accepted behaviour patterns. Famine can fling these gates wide open as tried and trusted mores fail to meet basic needs. The human instinct for self-preservation takes control. What determines the path any individual takes is a universe in itself.

Liz and I learned about culture shock at Bible college and wondered what would "shock" us in Niger. In Hausa fables and fairy tales we were surprised to discover that the villain—the cunning fox, the prowling hyena, the shrewd dealer who makes a profit at somebody else's expense, the deceitful and treacherous neighbour—was actually the hero!

Mammane always occupied a prominent position in village meetings, nodding and giving loud guttural clicks of approval. He was amongst the first to select and prune his trees—and always the first to show me. Like a wizened forester, he would rattle off a litany of benefits trees brought, bragging about his own exploits while denouncing others for not getting on board with the program.

Mammane maintained his trees till the very end of the 1988 food-for-work program but before the grain dust from the last food distribution had settled he removed every last tree from his field! As if shaking the dust from his feet, "finished with that nuisance Tony," Mammane was set free to do what he always wanted to do—get rid of those pesky trees on his farm!

The rains had failed but the tap roots of this kalgo tree drew water from deep in the soil profile, releasing some of it near the surface within reach of the roots of nearby crops.

Seini on the other hand, complained from the start, only reluctantly and belatedly doing the work. Seini was the eternal pessimist and was convinced that no good thing could possibly come from defying tradition. "How unreasonable of Tony," he would grumble, telling anybody within earshot—"doesn't he know that trees and crops don't mix? He is ruining our farms. What would our forefathers think?"

However, slowly, imperceptibly at first, a change began to occur. Weren't the millet plants growing near these young trees greener and taller than those further away? Weren't the trees pleasing to the eye? Wasn't it good to have even a little shade to put his water bottle and lunch under and to enjoy that ever so small but detectable cool air afforded by these aspiring trees? More to the point, Seini didn't have to worry about doing something different to his peers—everybody in the village was doing the same thing—leaving Tony's pesky trees in their fields in order to qualify for their monthly food ration! Seini felt safe in this space.

The food-for-work program gave Seini the opportunity to see what would happen if trees were allowed to grow together with his crops.

For Seini, his traditional village was his world. To be ostracized, ridiculed, singled out, was to court disaster. Who else did he have to fall back on in times of need but his peers? That brief period of famine relief opened the gates of Seini's mind just enough for him to imagine another possibility. It had created an environment in which it was safe to try something new. At harvest time, nobody was more amazed than Seini at the harvest. Not only did these emerging trees not reduce the grain yield, rather, a bumper harvest was realised! Seini's sentiment may well have been "well, we know Tony is crazy, but what he made us do didn't cause any harm and may have caused some good." At heart, Seini was a pragmatist. So far, there had been no downside, only benefits. Together with a few other enlightened but cautious farmers, Seini determined to persevere for at least another year, and see where this *bature's* (white man's) crazy ideas took them. And, when they reached the end of that next year, seeing there were no downsides again, only more benefits as the trees grew, they were emboldened to continue the practice for even longer.

Farmers listen to farmers

A Nigerien farmer once said to me, "Tony, if you tell us something, we don't really believe it, because if you are wrong, you will not suffer any consequences for your advice but we will. Even if your staff tell us, we do not believe them. They are paid to give that advice whether it is right or wrong. However, if a farmer tells us something, and we see him doing it himself, we listen because we know that his livelihood depends on it being right." He taught me an important lesson that I use to this day: farmers primarily learn from farmers. Farmers all over the world look at and learn from what their neighbours are doing. Peer-to-peer exchange is one of the most powerful ways to change beliefs and attitudes. Accordingly, in our FMNR promotion programs, a lot of weight is given to facilitating exchange visits, establishing model farms and villages and equipping and empowering farmers themselves to be agents of change.

Hassan

Hassan was married to Zeinu, a beautiful and energetic woman. They had lost three children in childbirth. In Hausa culture, being childless is a great source of shame for both wife and husband. Hassan was mocked as "the man with no heir" and Zeinu was not considered a grown woman. When Zeinu fell pregnant again, Hassan came to Liz and me for help. I organised for Zeinu to be examined at the highly respected SIM hospital in Galmi, some four hours' drive from their village. The doctor assured the couple that everything would be fine but it was critical that Zeinu return one month before the due date so that medical help would be on hand. We were to be away on holidays at that time and made arrangements for them to get to Galmi as agreed. When we returned from holidays, we learnt that the child was stillborn and Zeinu had bled to death in her village. We can only assume that Zeinu feared going to hospital and lied about her due date. While the situation has improved in recent years, in 2017, a staggering 509 mothers died per 100,000 live births in Niger.[32]

After his wife died in childbirth, I appointed Hassan to be my volunteer representative in his village. Hassan's efforts to ingratiate himself to me, earned him the nickname *Chef de Cabinet de Tony* (Tony's Chief of Staff). While he could talk the talk, he didn't actually believe in all this tree stuff. However, he was shrewd enough to know that I expected staff and volunteers to teach by example, and that sooner or later, I would want to see the trees on his own farm. So Hassan regenerated trees on a small corner of his land—the Just-In-Case-Tony-Visits Plot—and cleared the remainder of his land. The next year there was a severe locust plague. Hassan noticed that all of the millet on the cleared land had been devoured, while the millet in the plot with trees was untouched. Curious, he sat in the shade of one of the trees to try and understand why. As he sat there, he observed that every time a locust landed on a millet plant, a lizard would dash out from the tree, catch it, then dash back for cover. Breakthroughs in understanding and changing attitudes were sometimes as serendipitous as my own discovery of FMNR. Hassan became one of FMNR's strongest proponents.

Hashimou

Unlike most village chiefs, Hashimou was relatively young and open to new ideas and positive changes in his village. Leaders in tribal settings are often conservative and careful not to offend their power base. Traditional leaders hate being wrong or losing face. Hashimou didn't like this either, but he didn't let the possibility of losing face deter him from trying new things. Rather than being aloof, he fostered a certain camaraderie amongst his peers while still commanding the respect normally afforded his station in life. Rather than barking orders, Hashimou led by example, implementing all the project interventions in his own home and fields first, and then helping others implement the activities on their land. At the same time, Hashimou consulted with village members, outlining the situation they found themselves in and listing the likely consequences of doing nothing. He was very patient and would listen to everyone's point of view, even to the few troublemakers who were only ever interested in their own needs. I never heard him raise his voice or publicly humiliate anybody. Perhaps in doing this he also created a safe space for the adoption of innovation and change.

When I started going to the villages, I gravitated more and more to Hashimou and his village because of his openness and his leadership. He always brought others along in the work.

After a critical mass of farmers had become convinced of the value of FMNR, the biggest single factor in its spread was farmer-to-farmer exchange. Hashimou called village meetings,

Trees harbour natural predators of insect pests that attack crops.

gained consensus and organised working bees to accomplish the tasks at hand. Peer-group pressure can prevent people from doing things that they know would benefit them personally. The mindset may be "better to do without than to be ostracised." It may be that the unity of the community is cherished because it has been a powerful means of ensuring cohesion and survival in a harsh environment. In any case, having a critical mass of FMNR practitioners lent weight to the peer pressure against cutting down trees.

I worked in 100 villages, however, Sarkin Hatsi, was always my favourite and top-performing village. Hence it was the most likely place I would take visitors to—there was more to see, and visitors would get a good reception. Under Hashimou's leadership, Sarkin Hatsi was always among the early adopters of my new ideas. While there were many reasons to feel discouraged, Sarkin Hatsi was proof of what might be possible on a grander scale.

Hashimou remained important to me, as a friend, and as an agent of change, the whole time I was there. I got busier though and the time I could spend with him lessened as the project grew; but the bond was never broken.

Hearts and minds

Beliefs form the foundation for all decision making. Hence, my primary tools are not money, technology or expertise. Even though today I am sometimes called "The Forest Maker" and "The Tree Whisperer," I spend 95% of my time with people, not trees. We can only change landscapes by changing mindscapes. If I win that battle, the rest is relatively easy.

José Elías Sánchez understood this. He confronted false beliefs head on and helped people change their paradigm of themselves, their world and their God. They moved from negative defeatism to victory. He convinced peasant farmers that they were gifted people and that they could change their lot in life:

> It is like a battle over the hearts and minds of the people to drop wrong thinking about themselves, their community, their farms and God. And to adopt right thinking— that they are children of God, that working with others they can achieve much more than they can alone or against others, that their farms are capable of producing much, much more than at present and that God cares for them and loves them and is close at hand.[33]

No matter how skilfully crafted our project plan is, it is the ability to demonstrate a connection between FMNR and people's aspirations, sense of self-worth and dignity that will determine whether a "movement" is created or not. The spread of FMNR is not determined by governments or funds or NGOs; it's driven by the people themselves. What happened in Niger was not a technological or policy breakthrough and it did not rely on enormous injections of money. It was a people breakthrough. We will be successful at restoring vast areas of the world's degraded land when we learn to walk alongside those whose livelihoods depend on these areas. Together we can learn how to repair and maintain them, all the while restoring peoples' sense of pride and self-worth in the process.

Spontaneous adoption of FMNR elsewhere

Adoption of FMNR in some regions was spontaneous and not related to what we were doing in Maradi. In 1985, in the village of Aguié, farmers returning from work in Nigeria after the rains started had no time to slash the trees regrowing from stumps on their land. They noticed that when the crops of their neighbours were destroyed by severe windstorms, their own crops were protected by the "bushes" growing in their fields. Despite having planted several weeks later than their neighbours, they harvested their millet much earlier. After this experience, the practice of FMNR was adopted widely in the Aguié district.

Twenty years later with the support of the International Fund for Agricultural Development, 170 villages were involved in FMNR in this district and 53 village committees had been

Farmers continue to maintain tree cover to this day. The place where I discovered the underground forest in 1983 (left) and 34 years later in 2017 (right).

established, each encompassing three or four villages. Some 130,000 hectares of fields which were practically treeless in 1985 were covered with 103 to 122 trees per hectare under FMNR.[34] Today, Dan Saga (Son of Saga), which is closer to the Sahara and has a harsher climate, even sells firewood to neighbouring Nigeria.

In other places farmers were independently drawing their own conclusions about the importance of trees and were developing their own approaches to tree regeneration. For example, in Zinder Region, where a form of FMNR has a long history, farmers have almost exclusively promoted regrowth of one species—gao (*Faidherbia albida*), a fertiliser and fodder tree which is very compatible with Nigerien agropastoral systems. Legend has it that a Sultan of Zinder decreed that anybody who cut down a gao tree would forfeit their right hand—a strong incentive to comply. Some one million hectares of parklands dominated by this tree have emerged in recent decades.[35] Whereas gao is a component species in the FMNR practised in Maradi, other species more suitable for fuel wood and traditional construction such as kalgo (*Piliostigma reticulatum*) and sabarra (*Guiera senegalensis*, which are also found in Zinder) are dominant. These thornless species grow faster initially and have harder wood (better for fuel and construction) than *Faidherbia*.

While there is no doubt that our team had a big influence on the spread of FMNR across Niger, as these examples show, we cannot claim all the credit. First, FMNR in one form or other has been practised in Niger and beyond perhaps for millennia. Second, for anybody actively seeking to improve their land, FMNR is logical and intuitive. There are examples of similar practices in Honduras where Quesungual agroforestry systems are found.[36] During my travels I have come across numerous cases of farmers simply working it out for themselves. In the mountains of Aileu, East Timor, I met Manuel da Silva, an illiterate farmer who had never been outside his district. He was practising his own form of FMNR. His father had taught him to appreciate trees and during the struggle for independence from Indonesia (1975–1999), trees had protected his band of rebel fighters. When peace came, he began protecting the eroding hills on his farm by pruning the numerous "bushes" growing there. It is not important who developed and promoted FMNR. It is important that it is accessible and usable by farmers, independent of any government authority or external aid.

The restoration of hope

During my 17 years in Niger, the reforestation project received many visitors; including SIM administrative staff, representatives of our main donor, CIDA, and delegations from other NGOS.

Towards the end of our time in Niger, the parents of Canadian missionaries who were visiting asked to see our work. I took them to the village of Dan Indo (Son of Indo). Before going out to the fields, we sat in the shade of the village meeting tree. Gradually people came out and joined us. I asked them what was the most significant thing that the project had accomplished. I thought they would mention famine relief. After all, lives were saved and the assistance enabled them to stay home, and farm and harvest a crop the following season. Or perhaps it was the well we dug. Women used to spend hours each day drawing water from unlined wells in danger of collapsing at any time. Was it fuel-efficient stoves? *Zai* (compost) pits that could double crop yields? Fast-growing, drought-tolerant acacia trees with nutritious

Manuel da Silva was onto the "tree-stump thing" long before I visited East Timor in 2011.

edible seeds? Surely, FMNR? In just a few short years they had witnessed the restoration of a windswept, barren landscape. The risk of crop failure had been greatly reduced. There were now alternate income streams. Women could access firewood close at hand and more. To my surprise, they did not mention any of these accomplishments. They did not even mention FMNR at all.

I was disappointed. But what they went on to say touched me profoundly. They said, "Mr Tony, before this project, we were nothing, nobody. All we saw of our own district chief was the dust from his Land Rover as he sped through our village. The only time we saw a government forestry agent was when he came to fine us for cutting down trees. And all we knew of the outside world was what we heard on the BBC. Nobody knew or cared about us. Today, because of this project, we are known. We matter. Our district chief now calls on us for advice. The forestry agents hold us up as an example for all of Niger to follow. People from Canada, the USA, Australia, Europe come to see us to learn how we did it."

They had found their voice in the world. They had earned respect. They were proud of who they were. My nanna's desire for me to become a "somebody" has been fulfilled many times over in the lives of others!

While the impact of FMNR on the environment, livelihoods and food security was both expected and self-evident, in West Africa we noticed the emergence of something else that was just as, if not more, significant. I found people who were weighed down, living in uncertainty

"Mr Tony, before this project, we were nothing. We were nobody. Nobody knew or cared about us. But today, the forestry agents hold us up as an example for all of Niger to follow."

and fear of the future, people who had felt like hopeless victims of climate change and poverty. Now they were, to a large degree, masters of their own futures, and in a way that did not require ongoing dependence on foreign aid, expertise or goodwill. FMNR practitioners now had the power to address their own problems.

And with this, something very powerful was emerging. We began to see that the implementation of FMNR resulted in the tangible restoration of hope. Through the practice of FMNR and all that is involved in mobilising communities, communicating, planning, overcoming obstacles, working together towards a common goal, realising benefits and seeing changes in one's difficult circumstances, hope emerged.

People's dignity was restored before our eyes. Think of how soul-destroying it is for parents not to be able to adequately feed, clothe and educate their children. Year after year of poverty takes its toll, and too many farmers have low self-esteem and see no way out of their predicament. Hopelessness sets in. One of the greatest bonuses I received for my work was to witness a precious transformation as hope was restored. When people stopped being victims, and began to hope, they were changed forever. People were being empowered and liberated. FMNR does not create dependency. It is bottom-up development, putting individuals and communities firmly in the decision-making seat. When people comprehended the significance of FMNR there was often a spontaneous outburst of joy expressed through dancing, clapping, singing and laughing. They began to believe in themselves, gain confidence and plan for the future. People who have hope for a better future will send their children to school and invest in improving their land and farming techniques, which will enhance benefits even more.

When we first left Australia for Africa, we expected to return to Australia when Ben, our eldest, finished year nine, so that he could complete his final years of school in Australia. In May 1995, we were on leave in Australia and planned to return to Niger for just two more years. One day, Liz and I were talking. It seemed such a pity that, just when the work was starting to take off in Niger, we would not be there long enough to ensure it could continue without us. Soon after this conversation, during dinner, our kids said to us that they wanted to return to Niger for longer. Ben said that he was planning to study humanities so the transition back to the Australian education system would not be so challenging. Melissa added that, although

Melissa, Liz, Sarah, Ben, me and Daniel in 1995.

she knew that we needed to return to Australia when Ben finished school, she wanted to stay in Niger as long as possible, and Daniel and Sarah agreed. In July we returned to Niger for another four years.

While it was hard for our family to let us go in 1981, our departure did not mean a severance of relationships. Expensive phone calls were reserved for birthdays and Christmas but sending aerograms became a weekly routine. Letters were eagerly received at both ends and many remain in the family archives. We received visitors with great joy. Liz's parents David and Jill, my brother Peter, and several friends did manage to come to see the project and to spend time with us.

During our visits to Australia, I was moved to see that Dad had a stack of our SIM support brochures on the customer service desk in his garage. He would introduce me to customers and explain what I was doing in Africa. Dad loved to see our photos and hear our stories. He was very relieved that we were safe, and as the work unfolded, became very proud of us. Sadly, despite planning to visit, Dad never made it to Niger. He passed away in October 1995 following a brief illness.

As we approached the end of those four years, it became clear that it was the right time to leave. Wherever I conducted famine relief I had become well-known and was constantly bombarded with requests for help. It became so demanding that I had to employ someone to screen visitors at my office door. Dealing with so many people left little time to do more strategic work. At the market I would be quickly surrounded by onlookers so that even something as simple as buying food became unbearable. Despite having taught villagers how to be self-sufficient, they continued to depend on me. I felt that if I left, people would no longer be able to look to me for help when they had a problem. I wanted them to evaluate for themselves the various interventions which the project had introduced—such as growing hardy, food-producing shrubs and trees, fuel-efficient stoves, composting and mulching, and, of course, FMNR—and choose what they wanted to implement on the basis of direct benefit, and not because they wanted to please me.

Other factors converged to reinforce our decision to leave. Armed bandits made the road to and from our children's boarding school in Nigeria too dangerous to travel. Our house-help was getting married and Liz had a strong sense that her sewing machine would make a suitable wedding present. It would be a symbol of our parting and letting go of our life in Niger. So, we decided to return to Australia in 1999.

Leaving Niger

Amidst the busyness of preparing for handover—packing up belongings, dispensing with what we no longer needed, and attending farewell celebrations—I made a final visit to the villages and district government offices to say my goodbyes. The name of the newly installed *Sous-préfet* of Madarounfa district sounded familiar. I soon discovered that he was none other than the former Nigerien consul in Kano, the person who had expedited our visas 17 years earlier! It turned out that the consul was from a local royal family in the district I had served in. I seized the moment and recounted to him what his actions had meant—that we were able to work with his people to fight desertification and poverty, dig wells, provide food aid, and improve agricultural and health practices. He was quick to express gratitude for the ways we had helped his kinsmen, community and the district he now governed. "When you do good in the world, it will come back to you."

CHAPTER ELEVEN
A global movement

It was never going to be easy to leave Niger. I certainly loved Australia, and it was time to return, but I couldn't get excited about an office job. Whereas I felt called to go to Africa and all life's guideposts had pointed in that direction, I didn't feel called back to Australia. Still, I knew it was the right time to return. Much prayer and soul searching went into our decision to come home. When my inner turmoil was at its most intense, a friend sent me an advertisement for the position of Program Officer with the international Christian relief and development organisation, World Vision Australia (WVA).[37] I immediately felt that I could wholeheartedly throw my energy into this role. It was an unprecedented opportunity to apply on a larger canvas the hard lessons I had learnt in Niger. I would have the chance to spread FMNR to the world. Yet, it's just as well we can't see into the future. If I knew how long it would take to accomplish something and what the setbacks would be, I might never have started.

My interview for the role at WVA was conducted by telephone. Later, in a return call I was informed that I had the job. In my excitement, I picked up Liz and Melissa, one under each arm, and swung them around, inadvertently cracking one of Liz's ribs just weeks before we were to fly home!

Apart from home assignment in Australia every four years, we lived in Niger for 17 years before permanently returning to Melbourne where our children finished their schooling and attended university. In our first year back, we lived in a rented house before buying our own home, conveniently close to a school, shops, church and work in the eastern suburbs of Melbourne. Whereas I've never felt at home surrounded by buildings and traffic, Liz, who grew up in Melbourne, made the transition much more easily. Liz decided not to go straight back to work and

Visiting the Humbo community managed reforestation project with my World Vision Ethiopia colleague, Kebede Regassa.

Through repetitive drawing of water, ropes wear deep grooves into the hardwood logs placed at the rim of the well.

became the family anchor point as the rest of us found our bearings.

For several years after returning home, barely a day would go by without some small or large incident causing me to reflect on my time in Niger. A tap made me think of water. Running water is so apparently inexhaustible, clean and convenient in urban Australia, but scarce, contaminated and arduous to collect and carry in much of the developing world. In Niger there are villages where women queue up most of the night waiting for water to seep into the well. I remembered the hardwood logs (*Prosopis africana*) laid across the open mouth of village wells. The deep grooves in the logs were silent testimony to the daily passage of hundreds of hand-plaited ropes made from palm leaves (*Hyphaene thebaica*).

Even though I was back in Australia, I could not squander this precious resource. To this day, I collect the shower and washing water from the house and carry it to my garden. In any case, water abundance in Australia, the driest continent on earth, is only an illusion. As a nation, we need to learn to use it more wisely. Excessive extraction during dry years has contributed to insufficient environmental water flows, major fish kills and excessive salinity in the lower reaches of the Murray–Darling river system in south-eastern Australia.

Our experiences in Africa shaped all our children. Melissa, Daniel and Sarah settled into a local Christian school and were soon involved in sporting and church activities. Ben found casual work while preparing to go to university to study international development. Each one made the transition in his or her own way. They responded to the challenge with anything from resignation to excitement. To being wrenched out of Africa, their reaction ranged from indifference to rage. Our friends commented on how well we were adjusting to life back in Australia. They didn't see our kids' frustration when teachers told them how good it must feel to be back "home"; or when others insisted that they betray their passion for soccer and choose an Australian rules football team; or when their peers were so focussed on fashion and mobile phones, which our kids saw as trivial. You can take a person out of Africa, but you can't take Africa out of the person. It was challenging to predict the right clothes to wear in Melbourne, famous for hosting four seasons in one day. One of the hardest adjustments for them was the change from being well known and part of a community working to make significant change in West Africa to being anonymous amongst their peers in Australia.

Each of our children has come away with a great appreciation and respect for other cultures, a concern for people less fortunate than themselves and a heart for serving others. Ben completed a degree in International Development Studies, and now works in the mental health sector. Melissa has a bachelor's degree in nursing and a master's degree in divinity. She works part-time as a nurse. Daniel completed theological studies and works with university students, and Sarah completed a degree in outdoor education and teaches at a Christian school. All have married and we are just three wonderful grandchildren short of producing a soccer team.

We were surprised at how much time and energy was required to help our children make the transition and adapt to life in Australia and were glad that Liz was available to support them full-time. However, over time, Liz became frustrated with not being part of something bigger. Doing some computer courses at community training centres paved the way for future opportunities. One day in 2002, I rang her at home and asked if she could come to the office. I had ridden my bike to work and had forgotten my good shoes which I needed for an important meeting. After delivering the shoes, she bumped into Sheldon Rankin, the head of international programs. Sheldon had just been talking to a mutual friend and Liz's name was mentioned in their discussion about how the talent and experience of returned missionary wives was often wasted. On hearing that Liz was interested in getting back into international development, he invited her in for an interview which resulted in her full-time employment and a very satisfying six-and-a-half years in World Vision Australia's Asia, Pacific and Latin America teams.

If ever there was an organisation equipped to transfer an appropriate development approach to millions of people in the developing world, it is World Vision. I went through a steep learning curve with the dizzying multitude of acronyms, development terms and compliance requirements. Even though I had been a development practitioner for 17 years in Niger, there was much I didn't know about current thinking and donor requirements. I had to get used to attending many meetings, monitoring projects while working remotely and communicating by email. My desk overlooked Springvale Road where 60,000 cars passed by each day. My head spun and at times I barely held myself together. In Niger, I was used to open spaces and being able to make my own decisions and give directives. But, if I looked past the traffic and suburbia, I could see the forested Dandenong Ranges, a series of low mountains about 35 kilometres east of Melbourne's city centre, and somehow this view gave me patience to persevere.

Humbo, 1999

In my role as a program officer in the Africa team, I was initially responsible for projects in Ethiopia and Kenya. My first field visit was to Humbo, Ethiopia, to monitor the Likimse Water and Sanitation project. Humbo is located 320 kilometres southwest of the Ethiopian capital, Addis Ababa, and the project involved capping a large spring, building two 150,000-litre water storage tanks, laying 55 kilometres of pipe across rough terrain and installing 27 water access points to meet the needs of 40,510 people.

It's hard for us in the West to imagine the significance of such projects. On my travels I met villagers who had been hospitalised in order to have leeches removed from their throats; they had drunk contaminated water from stagnant pools. I saw girls drawing drinking water downstream from where cattle were standing and defecating. I visited a village previously renowned for its drunkenness, even among the children. The reason? The only way they could avoid diarrhoeal diseases was by drinking beer. The alcohol in the home-made brew killed the bacteria in the water.

I have the highest regard for my World Vision colleagues. As I engaged with them, I learnt about the sacrifices they made in the service of others. Many were stationed in remote areas with few modern conveniences, separated from their families for months at a time.

I remember a full day of hard driving with a newly married, young staff member to her posting where she would resume her responsibilities as a logistics officer. In 1999, long-distance phone calls were expensive. Her quiet tears laid bare the personal cost of her commitment to help the poor. The World Vision team often faced suspicion and opposition from the very people they came to help. They took prayer very seriously, rising early in the morning, and many fasted one day per week, praying for the people they served. In one new remote area program that I visited, while facility construction was underway, a one-room building served as the dormitory, office and dining room for the ten staff. The team consisted of people with qualifications in administration, engineering, health, agriculture, social science, and monitoring and evaluation. In the morning, beds were stood up against the walls and tables were brought out for meals, and then for work. They worked late into the night with flickering kerosene lamps. Even if they were tired, the team had to wait till everyone finished work so they could go to sleep. Often enough, that meant staying awake till past midnight.

With the return of the forest, dry springs began to flow again. Because less time was spent looking for water, children could attend school and women could tend to other tasks.

Green famine

In Niger, I always associated hunger years with drought and desiccated landscapes. So, in Humbo, I was shocked that despite the lush vegetation and tall grass, communities had experienced crop failure and World Vision was providing food aid. My Ethiopian colleagues explained, "This is green famine." I was intrigued. There had been enough rain in the growing season, but it fell unevenly and a dry spell at the critical grain-filling stage meant there was virtually no harvest. Additionally, the hills and farmlands were all but stripped bare of tree cover. When it did rain, the intense but brief bursts of subtropical rainfall resulted in flooding, causing loss of crops and sometimes livestock. As in Niger, deforestation and an over-reliance on just one or two annual crops made communities highly vulnerable.

What was happening in Humbo was a classic example of why biodiversity in both agricultural and natural landscapes is so critically important. While there are many purely environmental benefits, human beings also stand to benefit in multiple ways by restoring and managing biodiversity. Maximising biodiversity opens up the potential to benefit from the whole landscape, not merely the small plots. Likewise, it means production year-round, not only in the narrow window of time provided by an unreliable rainy season. Additionally, biodiversity enables farmers to continue being productive even in the face of environmental shocks such as drought, floods, severe storms and pest attack. In a diverse environment, there will always be some plant or animal that will either be immune to the shock, or which will be less impacted by it.

I could easily see how FMNR could help in situations like this, but I became discouraged from time to time at the slow pace of adoption.

An idea whose time had come

In January 2004, I had an opportunity to return to Niger. This was my first visit since we left in 1999. In a cafe in Niamey overlooking the Niger river I met Chris Reij, a human geographer from Vrije University, Netherlands. Chris had visited West Africa regularly for 30 years to document changes and the impact of agricultural innovations. He had just completed a cross-country survey and was surprised to see the extent of reforestation. In the 20 years since the tentative initiatives of individual farmers in 1984, FMNR had spread from farmer to farmer, and it was now practised on over half of the country's cultivated land. I was amazed and thrilled. World authorities had written off the Sahel as a lost cause. According to Chris, what had happened in Niger and neighbouring countries was the biggest positive environmental change in the Sahel, if not all of Africa. Chris slammed his pen on the metal tabletop. "Enough research! This story needs to be told."

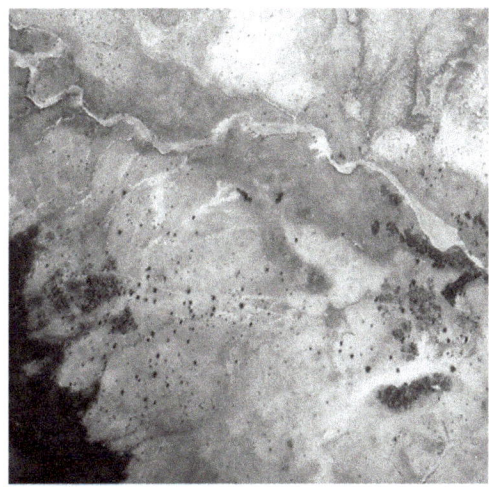 

Satellite photos showing tree density in the village of Galma, Niger, in 1975 and 2016. FMNR had exceeded my wildest expectations.

In September 2004, his findings were verified by Gray Tappan from the US Geological Survey. "I've been working for more than two decades in the Sahel and I've never seen anything like this." Gray's high-resolution, before-and-after satellite photos prove beyond doubt the remarkable scale and spread of FMNR in Niger. The satellite imagery revealed that there were about 5 million hectares of farmland in Niger with an average tree density of 40 trees per hectare. That's 200 million trees!

This was big news. A regreening FMNR movement, involving the systematic regeneration and management of farmer-selected, naturally occurring trees and shrubs from stumps, roots and seeds was spreading across Niger.

True to his word, Chris brought journalists and film-makers to the field and took FMNR champions to the world stage to tell their story. To attract funding for FMNR implementation, he worked with Dennis Garrity, UN Drylands Ambassador, as well as key figures in research, advocacy and policy such as Bob Winterbottom at the World Resources Institute. Chris deserves much of the credit for creating awareness for FMNR globally. A colleague asked Chris where he got the boldness to speak with heads of state, ambassadors and ministers. Chris channelled the Dutch football player, Dennis Bergkamp. "When you have the ball, and the goal is in front of you, you just shoot. You don't stop and think about it, test the wind direction, or consult others. You just shoot!"

Chris persuaded Lydia Polgreen, bureau chief for West Africa at *The New York Times*, to visit Niger and write up the story. On Sunday 11 February 2007, her front-page article "In Niger, trees and crops turn back the desert" hit the news-stands.[38] This feature caused a sensation in environmental and development circles. For the first time the transformation taking place in the Sahel was highlighted internationally.

Humbo, 2006. Community-Led Reforestation Project

World Vision had worked in Humbo since the 1984 famine crisis when it provided desperately needed food and medical aid. World Vision stayed on, moving into recovery and then development. The development phase involved establishing committees to make decisions and act on priority areas of need such as health, education, agriculture and economic development. Despite all the skilled and highly dedicated work, the population remained highly vulnerable. Such was the state of agriculture and food insecurity that for 22 years before the Humbo project's inception, to one degree or another, communities relied on food aid. This was due not so much to droughts as to widespread environmental degradation, reduction in biodiversity and over-reliance on a single, annual crop: maize. Of the 48,893 people living within the World Vision Humbo Program Development area in 2006, an estimated 85% lived in poverty.

The goal of the reforestation project was to regenerate 2,728 hectares of degraded forest and enhance local community livelihoods through improved environmental conditions, direct benefits from the restored forest and revenue from carbon credit sales. The project helped communities form cooperatives to invest revenue wisely. Construction of grain storage facilities meant farmers were no longer obliged to sell grain at harvest time when prices were low. When their own food supplies ran out, they could purchase grain from

> The cooperative in Humbo, Ethiopia, used the income from the sale of carbon credits to build a shed and buy a flour mill, saving the women hours of walking and waiting for their grain to be ground in distant market towns.

the cooperative at less than market price. Purchase and installation of grain mills in villages has reduced the burden on women who previously walked long distances to have their grain milled.

Amazingly, only six years after project inception, the seven previously aid-dependent communities managing the Humbo forest sold 106.7 tonnes of grain to the World Food Program (WFP). And, in 2016, when much of Ethiopia was ravaged by drought and severe food shortages, communities in Humbo were food secure. Farmers told me, "We received less rain this year, but whenever clouds hit our mountain, they were more likely to drop their rain on our land than on our neighbours' land." What rain did fall was more effective because temperatures and wind speeds were lower, and with more organic matter in the soil, more moisture was retained and less was lost to run off and evaporation. In one stroke, this land restoration project, backed by a steady income from sale of carbon credits, was and still is, restoring the environment and increasing incomes and resilience to climate change.

The success and impact of this project is testimony to the skill and hard work of World Vision Ethiopia colleagues who also became close friends. Assefa Tofu, Hailu Tefera and many others put in countless hours, and they were sometimes tested to the limits by fearful villagers who thought they would lose their access to forest resources and, at times, by officials who questioned the whole venture.

Measuring the success of the Humbo project in economic, environmental or climate terms alone fails to capture the full impact. In 2012 Anna Szava, an Australian PhD student, recorded community members' responses to the changes:

We are too-much-happy. I feel only happiness when I look at the hill.

"When we walked to Hobicha market there was no shade to rest in, but now we walk in the shade of trees the whole way!"

When there was bare rock we did not have rain. Now there are trees, we have rain.

Today I feel very proud. I feel proud not only for myself but for my village and the district. This project is known in the whole world.

The bare hills of Humbo, Ethiopia, experienced a rapid transformation following adoption of FMNR by surrounding communities.

This kind of enthusiasm is infectious. The late Yamdaan Zimbil Longmoate, chief of Yameriga, Upper East Region of Ghana, declared that "this gift of FMNR is from the almighty God and therefore anywhere you visit you bring life and joy." Life, joy, hope! This is the stuff of movements!

Experts are perhaps the most difficult to convince. The smartest minds, millions of dollars and much rhetoric have been directed to reversing desertification, but in all too many instances, to little avail. The very things which have made FMNR successful have been stumbling blocks to some. Because FMNR is low cost, it has been perceived as cheap and second rate. Because it is farmer managed, it has been interpreted as unscientific, backward and haphazard. Because the species of trees we regenerated were indigenous, FMNR was often seen as wild, indiscriminate and inferior. However, these shallow perceptions about FMNR quickly melted away; not just because of the significant empirical impacts, but more compellingly because of the enthusiasm and energy of those directly affected. People liberated from debilitating poverty and hunger now had agency to create a better future for their families and become powerful advocates. Thousands of Ethiopian and international visitors including government and community leaders, NGO practitioners, donors, journalists, researchers and ordinary farmers have come away inspired by what they saw and heard. In time, Humbo became a catalyst for the spread of FMNR through the World Vision partnership and beyond.

The Humbo reforestation project is a certified Gold Standard[39] carbon sequestration project under the Clean Development Mechanism.[40]

It involved the World Bank, the Ethiopian government, the Humbo community and World Vision. As the first of its kind in Africa, the third largest in the world and World Vision's first forest-carbon trading project, Humbo generated a lot of interest. Today, World Vision plays a lesser role and the community manages the activities itself, while carbon is sold on the voluntary market.

FMNR Hub

Despite a growing number of success stories, progress in spreading FMNR to other countries was slow. In 2012, a colleague, Rob Francis became World Vision Australia's FMNR project manager. Rob saw the simplicity and impact of FMNR and how I was struggling to get the message out to more people. Rob conceived the idea of establishing the FMNR Hub and website,[41] as a virtual centre of excellence. The Hub promotes FMNR globally through coordination, communication, collaboration, technical support, building scientific credibility, advocacy and project fundraising. Rob helped raise funds allowing me to take FMNR to the world. The budget covered travel expenses enabling me to create awareness and give face-to-face training in countries I would otherwise never have reached. The Hub gave a platform for advocacy, fundraising and for supporting fledgling work and struggling FMNR promoters in the field. Without this timely and insightful help, FMNR would not have spread at the rate it has. For the first time I had the freedom and the funding to proactively take FMNR to the World Vision partnership and to the world.

Various other World Vision colleagues, too many to name, made significant contributions to the work through their endorsement, project evaluations, development of the FMNR Project Model which guides proposal development, creation of an online training course,[42] compilation of the FMNR manual,[43] development of monitoring tools, fundraising and media campaigns, and integration into World Vision strategies and working plans. Impact in the field in turn resulted in widespread recognition of the work through awards, greater media coverage and further awareness and uptake.

2012 Beating Famine Conference

Meeting with journalists during the FMNR conference in Ghana, 2009.

In Niger, FMNR was first promoted successfully at village level. With a slight increase in effort, it was promoted at district and national levels, still having good impacts. I wondered what it would take to promote FMNR at regional and international levels? Over the years, I've found that an incremental increase in effort can result in an exponential increase in the rate that FMNR spread. While my preference is to keep out of the limelight and work at village level, it seemed a natural next step to work with others in large public forums, bringing FMNR to a global arena.

Again, Rob Francis at WVA stepped up to the challenge and along with the indomitable Dennis Garrity and a small team of dedicated support staff, developed and organised the first Beating Famine Conference in Nairobi in April 2012.[44] In partnership with the World Agroforestry Centre and the Evergreen Agriculture Partnership, World Vision hosted this international conference in Nairobi to analyse and plan how to improve food security for the world's poor through FMNR and Evergreen Agriculture,[45] an initiative to integrate trees, food crops and livestock for more sustainable and productive agricultural systems for smallholder farming families.

The Beating Famine Conference was attended by more than 200 participants, including world leaders in sustainable agriculture, five East African ministers of agriculture and the environment, ambassadors and other government representatives from Africa, Europe and Australia, and leaders from non-government and international organisations such as the Food and Agriculture Organization of the United Nations (FAO). Of significance for World Vision, Tim Costello, CEO of World Vision Australia at the time and a great supporter of FMNR, gave an opening address and prayer and participated in the conference for its three-day duration.

The Beating Famine Conference planted the seed for a global network of key stakeholders. For the first time ever, plans for FMNR scale-up were made at country, regional and global level. A trend of regular media coverage in some of the world's leading outlets commenced, and a noticeable increase in momentum for an FMNR global movement was triggered.[46]

On the road together with Liz

From 1999 to 2012, I travelled frequently while Liz held the fort at home. She supported the kids, chauffeured them to and from sport, church and social commitments, taught all four to drive, smoothed the transitions from school to tertiary education, met their future partners and organised weddings. She also worked in schools, at World Vision and with the Victorian Roads Authority (VicRoads). But her job description was to change.

With the establishment of the FMNR Hub, there would be much greater demands on me to travel. For the sake of our relationship, Liz and I felt that either I had to pull back—at the very time all the years of hard work put into promoting FMNR were about to bear fruit—or Liz should travel with me. In 2012 she took leave from her work as an Executive Assistant with VicRoads and joined me for three months in East Africa. Together we helped facilitate the inaugural Beating Famine Conference in Nairobi and workshops and training events in Kenya, Ethiopia, Tanzania and Uganda. I love my work and find fieldwork exhilarating and invigorating. On the flight home Liz looked elated and restless. "What have you done to me? How can I go back to an office job after this experience?"

Liz found her work at VicRoads stimulating and satisfying and she was appreciated and popular, but the trip had rekindled her passion. I was overjoyed. Once again, we were united in our calling and vision. Liz left her job and for five years travelled with me for several months each year. She played a crucial role in networking and supporting me and the local World

"This gift of FMNR is from the almighty God and therefore anywhere you visit you bring life and joy." Yamdaan Zimbil Longmoate, chief of Yameriga, Upper East Region of Ghana.

Vision staff running workshops, training trainers and facilitating national conferences.

This hectic time of planning and travelling, living out of suitcases and relying on each other has been among the happiest and most fulfilling periods of our lives. Together we engaged in the ambitious undertaking of inspiring others to restore their landscapes. We often had the privilege of revisiting farmers—two or three years down the track—to share the joy of their accomplishments. We witnessed their "aha" moments—when failure and defeat are replaced by a deeper comprehension of their problems and a clear path forward. Along the way we have met so many inspiring people and felt privileged to play a small role in their unfolding stories. This "aha" moment has proven to be an ignition, a point of no return for many individuals who went on to have huge local impacts. Some even generated their own FMNR movements.

Movements, not projects

Liz and I did not want to simply duplicate FMNR projects in various countries. The time is too short and the area too great to rely on the simple addition or even multiplication of projects. Conventional projects have defined start and end dates, set geographical foci, expected outcomes and fixed budgets. Frequently, once the last dollar is spent, the work stops. As far as possible, we wanted to replicate what happened in Niger. Unlike projects, social movements do not have clearly defined start and end dates. They are spontaneous. They have the element of surprise. Movements are not limited to geographical boundaries, their outcomes can exceed expectations, and their outputs are not necessarily proportional to the budget.

The work spreads without funding because movements have their own momentum and energy. In Niger, once a critical mass of people adopted FMNR, the self-evident benefits and elimination of obstacles propelled adoption to new levels.

Social movements by their very nature cannot be orchestrated. We can implement key elements which we think will contribute towards creating a movement, but, because movements depend on the actions of individuals and circumstances, we cannot predict when, or how a movement will occur. According to Malcolm Gladwell in his bestseller, *The Tipping Point: How Little Things Can Make a Big Difference*,[47] ideas, products, messages and behaviours spread just like viruses. Social epidemics are driven by a handful of exceptional individuals. Small actions can single-handedly initiate big and rapid changes.

Niger in 1984 was the right time and place for the emergence of an FMNR movement. The severe famine of 1984–1985 capped a series of hit-and-miss years in which farming practices were regularly failing to meet people's basic needs. Each year, people lived in greater uncertainty over growing sufficient food to see them through till the next harvest. In fact, it was more certain *not* to grow enough food than vice versa. The dire circumstances pushed people to change, while the compelling benefits of FMNR pulled people to change.

Further, FMNR is a "no regrets" technology. That is, there is little to lose through its application, but potentially very much to gain. For vulnerable people with few choices in life, this was one change that was easy to embrace.

FMNR can be applied over large areas of land and can be adapted to a range of land use systems. It is simple and can be adapted to each individual farmer's unique requirements. FMNR provides multiple benefits to people, livestock, crops and the environment. Through managing natural regeneration, farmers, pastoralists and forest-dependent communities can control their own resources without dependence on externally funded projects or the need to buy expensive inputs: seed, fertilisers and nursery supplies. Its beauty lies in its simplicity and accessibility to even the poorest. And once it has been accepted, it takes on a life of its own, spreading from farmer to farmer by word of mouth.

How do we create a successful FMNR movement in new settings with different circumstances? There is no single answer to this question. To best meet the needs of a specific location, at World Vision we draw from a suite of interventions. We create awareness through meetings, workshops, conferences and media. We create pilot areas to demonstrate the impact. We train and equip FMNR champions and facilitate exchange visits. We support farmers themselves as the premier agents of change and influence. We equip people with knowledge and skill on how to tackle obstacles to FMNR uptake. We engage leaders and all stakeholders (government, traditional and religious authorities, youth, women and marginalised groups). We create awareness of the link between deforestation and impoverishment and then we empower people to act. We walk alongside them for a period—visiting, learning together, coaching, correcting, encouraging and supporting, especially where there is conflict and danger of reversals. This is time-consuming. It requires patience and above all empathy; nobody cares about how much you know until they know how much you care. Successful implementation is about so much more than imparting mere technical knowledge and skills.

We encourage the formation of appropriate FMNR groups for solidarity and mutual support. Often there is a need for advocacy for laws to be changed so that farmers will have the necessary assurance that if they practise FMNR, they will benefit. We look for market linkages to generate economic sustainability and we facilitate value-adding through introducing or scaling up activities such as beekeeping and sale of fruit.

Having mixed all the known key ingredients into the recipe, we persevere, we continue to engage with farmers, and we wait—one, two, five years? In Niger, there were clear signs of emergence of a movement within 10 years, but we did not know the extent of the movement until 20 years later. Today, by building on the foundation already laid, we are beginning to see signs of movements in as little as one to five years. *Da yayyafi kogi kan cika*—Drop by drop, the river fills.

Bishop Simon Chiwanga, Mpwapwa, Tanzania

Retired Bishop, Simon Chiwanga from Mpwapwa, Tanzania knew the devastating impact deforestation and constant fires on the Kiboriani range were having on his parishioners.[48] He had tried unsuccessfully to close part of the range to human disturbance. In 2005, with the Jitume Foundation community groups, he planted two million trees, but most died and the community groups started to disintegrate. After attending an FMNR Workshop in Moshi, Tanzania, in March 2012 and the Beating Famine Conference in Nairobi in April, his newfound understanding rekindled his determination. He returned to Mpwapwa with renewed zeal to restore degraded farms and hills. He began on his own farm, then invited small groups of farmers to learn from his experience and visited their farms.

"I have been converted twice. First, when I met Jesus Christ and invited him into my life. And second, when I learned about FMNR and was given the power to restore the land and livelihood of my people." Bishop Simon Chiwanga and Gladys, Mpwapwa, Tanzania

> The tree stumps never cease to excite me, and I was delighted to find the same with others after the tree-stump thing dawns in their mind. Stumps were not quite visible where we began. When we got to a typical tree stump with a few shoots I could not resist my excitement. The group noticed my radiant face and asked for the reason. I explained the secret of FMNR—to release the underground forest to come to the surface, and that living stumps were the outlets for the underground forest to mushroom. A lady remarked, "Aha! Is that what we should be looking for? I have been doing a horrible thing burning tree stumps so that I could dig them out for firewood." We were standing in her farm. From then on, we were like game hunters, chasing living stumps.

In 2017, Bishop Chiwanga's own organisation, the LEAD Foundation,[49] began partnering with JustDiggit, a Dutch-based international NGO combatting land degradation and climate

change. Their ambitious program aims to vigorously scale up FMNR and very small-scale rainwater harvesting techniques throughout the Dodoma Region of Tanzania. By 2021 they expected half of the households of the Dodoma Region to restore 180,000 hectares of farmland with at least 14 million trees through FMNR, and 600 hectares of communal land across 300 communities.

Nineteen-year-old Austrian Josef Ertl was among the expatriate volunteers working on this program. After despairing of finding a solution to global deforestation and land degradation, Josef discovered the FMNR Hub. Josef wasted no time in enrolling in the online training course. To support grass roots FMNR movements and volunteers he created the Awaken Trees Foundation.[50]

From enemies to friends

Over the years I've learnt that if I can influence people's false beliefs and negative attitudes towards the environment, then destructive practices will be replaced by constructive ones. In 2013 I visited Mr Marimo Mbijima, a man in his mid-60s at his farm in Kongwa, Tanzania. He took me to the edge of his field and showed me a burnt-out tree stump with a new stem sprouting from the edge and said, "This tree used to be my enemy. I burnt the stump every year to get rid of it. Since learning about the value of trees it has become my friend and I am so ashamed that I ever tried to destroy it." That's when I realised that much of my work involves turning enemies into friends—enemies of trees, that is, into friends of trees.

> "This tree used to be my enemy. I burnt the stump every year to get rid of it. Since learning about FMNR, it has become my friend and I am so ashamed that I ever tried to destroy it."
> Mr Marimo Mbijima,
> Kongwa, Tanzania

More Beating Famine conferences and partnerships

Follow-up Beating Famine conferences held in Lilongwe (capital of Malawi) in 2015 and Bamako (capital of Mali) in 2019 built on the foundation laid in 2012 and further solidified FMNR's acceptance as a best-practice land restoration technique. This heightened awareness of

FMNR has created greater opportunities for worldwide spread. As a result, FMNR is now supported and promoted by an increasing number of international organisations.[51]

Roland Bunch recognised the power of FMNR. "I visited 385 agricultural programs in 95 countries. Very few programs have scaled up. The FMNR story is unique."[52] It is not only the wide range of impacts, its low cost or even the restoration of hope that makes FMNR different. It is its power to spread organically from person to person, with no external input that makes it stand out in the crowd of development interventions. We are now beginning to see FMNR movements unfold on a global scale.

Numerous influential articles produced by the World Resources Institute played a leading role in launching the AFR100 (African Forest Landscape Restoration Initiative) at the Paris Climate Summit (2015), with a target to restore 100 million hectares of degraded land in Africa by 2030.

"Thousands of projects have come through here, but with this one—if we are the judges—there is no comparison. The type of benefits we see pushes me sometimes to leave my home and walk through my field just to appreciate the trees."

In 2016, Gray Tappan published a study on a portion of Niger. The study estimated that FMNR was being practised on over 6 million hectares. A Niger government study estimates FMNR is being used on 10 million hectares. Either way, this is a significant achievement by the people of Niger. That such an environmental transformation happened in one of the world's poorest countries—in a physically and socially hostile environment with little investment in the forestry sector by either the government or NGOs—makes it doubly significant to countries facing similar problems. The Niger example raises the question: if such rapid and low-cost reforestation is possible under such difficult conditions, what might be possible elsewhere?

Indeed, the Niger story is a beacon of hope to a world suffering the twin scourges of land degradation and climate change.

In 2018, the Evergreen Agriculture Partnership was restructured, renamed the Global Evergreening Alliance[53] and registered as an international non-governmental organisation. The Alliance is attracting millions of dollars in funding for land restoration activities and FMNR is its flagship intervention.

FMNR movements: Kaffrine and Fatick, Senegal

Following a slow start, by 2015 FMNR had spread to 64,000 hectares in Kaffrine and Fatick in Senegal's peanut basin. During a project evaluation interview, Aissatou, a lead FMNR farmer said,

> Thousands of projects have come through here, but with this one—if we are the judges—there is no comparison. We have nothing but our environment. Since we started working with FMNR we have already started seeing the benefits that we have not seen with any other project. The type of benefits we see pushes me sometimes to leave my home and walk through my field just to appreciate the trees and environment. When things get to where they need to be, we will see more yields and the path will be clear.

This is a remarkable statement. Few farmers have time or motivation to visit their fields for the simple pleasure of appreciating the trees and the environment. Visiting one's farm is an obligatory duty if one is to eat and it involves great physical exertion, exposure to heat, privation and frequently disappointment brought on by drought and insect attack. A monumental shift in thinking and consciousness has occurred here. Thousands of individuals like Aissatou are quietly reaching out to their neighbours, freely and patiently giving their time, teaching, following up and leading by example to help others experience the same benefits.

I have had the privilege of knowing and working with many heroines of FMNR: from village level as with Aissatou; to project management level including design, monitoring and evaluation; and management of international programs. Their unique insights, strength of character and skilled leadership has propelled FMNR to new heights.

FMNR movements: Upper East Region, Ghana

In Ghana I had the pleasure of working with the late Norbert Akolbila on the Talensi Area Development Program in the Upper East Region. In December 2014 he resigned from World Vision Ghana for health reasons and to follow his new passion—FMNR. In his resignation speech, Norbert explained:

> After the FMNR "baptism" I received from Tony in July 2009, let it not surprise you to find me after World Vision fully engaged in the FMNR movement. No matter where I find myself, I can assure you that FMNR is something I cannot part with for the rest of my working life.

"No matter what the season or circumstance, a diversity of crops means there is always something to eat and something to sell. FMNR has given me stability, abundance and resilience."
Yaouza Harouna, Maradi, Niger

He went on to found and lead the Movement for Natural Regeneration (MONAR), the world's only NGO solely dedicated to the spread of FMNR. MONAR had no funding, so Norbert funded MONAR's activities himself. In its brief operating period, MONAR touched the lives of thousands of people. Sadly, on 12 February 2017, Norbert passed away. I lost a good friend and one of the most effective agents for change I have ever met. Norbert did not work in vain. Josef Ertl's foundation, Awaken Trees, is funding the revamped NGO, now named Forum for Natural Regeneration, ably headed by Sumaila Saaka.

Apple of the Sahel

In 2017, I interviewed an energetic 39-year-old Nigerien farmer, Yaouza Harouna. Yaouza is married and has six children. Amongst the trees he regenerated on his three hectare, rainfed farm were kurna (*Ziziphus mauritiana*), a thorny, fast-growing tree which produces edible fruit of low commercial value. He added value by grafting improved Ziziphus fruiting varieties called "apple of the Sahel" onto his wild trees. He also raises free-range chickens in his fields. Yaouza told me:

> This year the raining season was not good enough to produce our cereal crop, but I earned not less than US$400 from the sale of Sahel apples alone. With this money, I bought three sheep at $200 which I recently sold at $790 giving a profit of $590. With this I was able to meet household needs: food provisions, children's education, health, support for relatives, and I even bought a motorbike. To protect my orchard from theft, every year I hire a guard at $80 per season to stay on the farm. Since adopting FMNR with Sahel apple grafting I have peace of mind.

CHAPTER TWELVE
One billion hectares

Silvia Holten, Media Director for World Vision Germany, had long insisted that for the sake of the world's children, World Vision must deal with the issue of climate change. In early 2012, a journalist from the daily newspaper, *Die Welt*, asked her about the Great Green Wall.[54] Despite being a relentless climate action advocate, Silvia was unaware of World Vision's reforestation work and unable to respond. She was surprised when the journalist told her that reforestation was World Vision's primary response to climate change. Silvia's research soon led her to FMNR and my story. "I was electrified, because it was immediately clear to me that the FMNR method marked a turning point in the climate debate. After I met you in 2012 at the Beating Famine Conference in Nairobi, I lectured everyone at World Vision about FMNR, whether they wanted to hear it or not!"

Silvia never goes anywhere alone. Following the 2012 Beating Famine Conference in Nairobi, she invited a journalist Horand Knaup from *Der Spiegel*,[55] Maria Ruether from the German NGO umbrella organisation Aktion Deutschland Hilft (ADH) and a filmmaker to visit the Humbo reforestation project. Such visits were to be repeated with dignitaries and high-profile media representatives in the following years. News of transformation captured the imagination of audiences so used to hearing bad news about the environment and climate.

Back in Germany, Silvia was contacted by consultant Martin Falkenburg, from ZAR University in Bielefeld. He was interested in fighting climate change and growing global poverty, and he wanted to support a project that could have a major impact on our future. When he came across my work on the internet, Martin became so enthusiastic about the topic that he immediately contacted Silvia, wanting to support World Vision and

FMNR makes people clap, sing and dance.

"FMNR is the answer!" Silvia Holten, Media Director for World Vision Germany, gave me a voice and a global platform.

promote FMNR. With Silvia's help he founded an FMNR regional group, Regionalgruppe Bielefeld FMNR, in 2014.

As they thought further about what they could do together to promote FMNR they came up with the idea to nominate me for the prestigious Right Livelihood award, also known as the Alternative Nobel Prize.[56]

In the meantime, I was experiencing increased levels of frustration. On the one hand, there was growing interest in FMNR in Europe and the USA, and World Vision field offices were recognising the high impact of FMNR interventions. On the other hand, a global decline in giving to charities made it more difficult to be proactive in taking FMNR to new regions such as Latin America and to support fledgling efforts in countries where it was being introduced, such as India and Myanmar. While we were entering a period of gathering strength and momentum in countries where FMNR had been introduced, there was stagnation in terms of global spread. This was a bitter pill to swallow. So much of the foundation for growth had been laid, so much momentum had been gained, but a successful global movement still seemed out of reach.

In 2012, I set myself what at first seemed the over-ambitious goal of introducing FMNR into 100 countries by 2030. In 2018, my nomination for the Right Livelihood Award had been submitted for the third time. The fact that the Award committee had encouraged Dr Falkenberg to resubmit the nomination three years running made me suspect I was being given serious consideration. While boarding a flight for East Timor, I noticed a missed call with a +46 country code. I googled the number. A call from Sweden could only mean one thing!

My fellow 2018 Right Livelihood Award Laureates (from left to right): Thelma Aldana, Yacouba Sawadogo, Iván Velásquez and (pictured) Abdullah al-Hamid, Waleed Abu al-Khair and Mohammad Fahad al-Qahtani who are serving lengthy prison sentences.

While not a complete surprise, I was thrilled and excited. I also felt vindicated. Surely now the message about FMNR would be heard and heeded.

The Right Livelihood Award was established in 1980 by the Swedish-German philanthropist and stamp collector Jakob von Uexkull to "honour and support those offering practical and exemplary answers to the most urgent challenges facing us today." An international jury decides the awards in such fields as environmental protection, human rights, sustainable development, health, education and peace. It has become widely known as the "Alternative Nobel Prize" and there are now 186 Laureates from 73 countries. It is one of the most prestigious awards in sustainability, social justice and peace.

The 2018 ceremony was the 39th Right Livelihood Award Presentation and was held in the Vasa Museum in Stockholm. During the ceremony Thelma Aldana (Guatemala) and Iván Velásquez (Colombia) received Honorary Awards "for their innovative work in exposing abuse of power and prosecuting corruption, thus rebuilding people's trust in public institutions." Farmer Yacouba Sawadogo (Burkina Faso), was recognised by the Award jury "for turning barren land into forest and demonstrating how farmers can regenerate their soil with innovative use of indigenous and local knowledge." Regrettably, the three Saudi Laureates (Abdullah al-Hamid, Waleed Abu al-Khair and Mohammad Fahad al-Qahtani) were prevented from attending due to lengthy prison sentences for their work promoting justice and equality. I received the Award "for demonstrating on a large scale how drylands can be greened at minimal cost, improving the livelihoods of millions of people."

Ole von Uexkull, Executive Director of Right Livelihood, said: "The Laureates' trailblazing work for accountability, democracy and the regeneration of degraded land gives tremendous hope and deserves the world's highest attention. At a time of alarming environmental decline and failing political leadership, they show the way forward into a very different future."

Receiving this award changed my life. The award has catapulted FMNR onto a world stage and greatly contributed to its wider acceptance as an effective tool in land restoration amongst donors, policymakers and development practitioners globally. I am regularly asked for an interview, an article or opinion piece, to appear at major international conferences or to speak in clubs, schools and churches. The award has given me an audience with ministers of development, captains of industry, religious leaders and secretaries of state. It has given me an opportunity to advocate on behalf of the poor and the environment. Perhaps my goal is not beyond reach after all?

I never set out to receive awards or gain recognition. They have simply been one of the bonuses along the way. It still feels odd when strangers want a selfie with me! My greatest reward occurs every time I go back to an area where landscapes and people's lives have been transformed. Seeing such joy and pride, children thriving, a surplus of food and hope in people's faces is reward enough. I have lived a blessed life to play a small role in these changes.

From vicious cycle to virtuous cycle

The Senegalese have a proverb which says: "If you are on a journey and lose your way, go back to where you started, and from there you will be able to work out the right direction." Farmers and land users who watch their land degrade know that they have lost their way. They know that their farming practices do not meet their family's needs and are even less likely to do so in the future. While not all traditional farming methods were in harmony with nature, there was greater utilisation of biodiversity. There were rotations and recovery (fallow) times which allowed the land to heal itself. However, as populations have grown there has been greater pressure on land. Fallow is practised rarely or not at all. With modernisation, the trend is to remove all trees from farms and rely on a few, usually annual, staple crops. FMNR offers a way for farmers to return to where farming started, in the forests and on treed plains.[57] From there they can work out a better direction to take.

In March 2019, I returned to Niger with Oscar-award-winning filmmaker, Volker Schlöndorff, to make a documentary on the FMNR story. In Sarkin Hatsi, I was reunited with Dan Mani. His affection was overwhelming.

"Welcome, welcome, Tony, is it you? You came! Tony has come! Praise God! How was your trip? How has your year been? Is Liz well? How is Ben, the boy who used to catch our goats, and climb our compound walls? What does he do now? Is he married?"

The people in the villages never forgot Ben, and on each visit they ask after him.

Dan Mani was an early adopter of FMNR and we spent many hours together. One hot day after coming back from the field with him I took my shoes and socks off. His young son exclaimed "Father! Father! Even his toes are white!"

Dan Mani and a small crowd from the village proudly showed me their fields. They recounted all the things I had taught them—things which seemed foolish to them at the time. Now they wanted so much to show how they had adopted them and were benefiting from them. Perhaps the most surprising change was that they no longer burnt the crop residue. After testing the things which I told them, they soon realised the powerful fertilising and water-holding characteristics of organic matter, and stopped burning altogether.

I was deeply moved by what I saw. Through FMNR-led restoration of trees into the landscape, the vicious cycle of destruction and poverty that I witnessed in 1981 had been replaced by a virtuous cycle of restoration, increased soil fertility, diversity and greater prosperity. Old friends with whom I worked in the '80s and '90s eagerly took me to their fields to show me their work. Not only had they continued managing the trees, but other practices which they had not adopted while I was with them were now standard. The younger generation had never met me, but the stories of fighting deserts and hunger had been passed down by their parents In each village we visited, children and youth sat patiently to see *Malam* Tony for themselves.

More trees in the landscape and less use of fire made more fodder available, which in turn enabled increased livestock numbers. This even included free-range poultry, not only in the village, but in the fields, which now had an assortment of insects to forage. In the past, it was rare to see poultry outside the village boundaries during the dry season. More livestock (along with increased tree-leaf litter fall and incorporation of crop residues) has led to improved soil fertility and greater soil moisture-holding capacity. This in turn has resulted in higher grain yields, and increased resilience against crop failure even in years of lower rainfall and seasons with long dry spells in between rains. More tree cover has resulted in less wind damage and less soil erosion. Food security appears to have increased, as have incomes. Farmers now have more diversified income streams: fodder, livestock, wood for fuel and poles used for traditional construction. FMNR has also provided a foundation for agricultural diversification, with some farmers greatly increasing their income from honey, grafted fruit trees, poultry and livestock, and others reintroducing sesame, cassava and sweet potato, which have not been grown in many villages for decades. This diversification has increased resilience and reduced vulnerability to climatic shocks.

My joy was tempered by some other changes: many of my dear friends, who had struggled with me in the 1980s and '90s, including Hashimou, had passed away. I felt Hashimou's loss very deeply. He had always been there with a smile—open, listening, quietly leading. And now he wasn't. An era had passed.

Security had become an issue as armed thieves based in remnant forests near the border with Nigeria stole goods and kidnapped passers-by for a ransom. To my dismay, Islamist extremism had entered Niger and continues to menace innocent villagers to this day. I was visiting South-Central Niger which is relatively calm and distant from Boko Haram (concentrated in the east) and Al Qaeda-affiliated groups based in Mali (in the west). However, no part of Niger is beyond their reach and rumours of their presence circulated. These threats necessitated an armed escort to the villages that I had visited freely in the past. Finally, wherever we went, we observed vast numbers of children and

"With trees, our land breathes again!"
Where it all started: the village of Waye Kaye and its chief, Dan Lamso.

youth. I wondered where they would all find meaningful work or any work at all. Would their sheer numbers negate the great gains that had been made?

Today, the politics of Niger take place in a framework of a semi-presidential, representative, multi-party, democratic republic, whereby the President of Niger is head of state and the Prime Minister of Niger head of government. The government is committed to tackling issues of poverty and food insecurity; however, in my view, its necessary war on terror has reduced its ability to focus on these challenges.

No time to waste

At the present time there is more to do than ever. By 2050 the world's population will grow by over 20% to 9.8 billion.[58] Nearly all of this population increase will occur in developing countries. Even if we reduce food waste and stop feeding grain to livestock, food production will need to be stepped up considerably. This is happening at a time when good agricultural land is being lost to degradation: every year, the world loses 24 billion tonnes of fertile soil.[59] Climate change is making productive areas more marginal and making it more difficult to farm. Twenty-six million hectares of forests are still being lost each year[60] and biodiversity continues to plummet with one million species at risk.[61] The rate of destruction is simply breath-taking. If we are going to feed the world while retaining any quality of life and not live in a completely artificial environment, we will have to do so in harmony with natural processes.

In 2019, the Intergovernmental Panel on Climate Change gave the world 12 short years to cut carbon pollution by 45% by 2030 and to achieve net-zero emissions by 2050. This is necessary in order to keep average global temperature rise to a maximum of 1.5 degrees Celsius and to avert extreme heat, drought, floods and poverty. Judging by the lack of response by some governments, the report might as well have been written on hot air balloons. However, to those of us who take these warnings seriously, there is no time to waste. The UN Secretary General termed the August 2021 IPCC report a "code red for humanity."[62] "The original ambition of limiting global temperature rise to 1.5°C is already 'perilously close' and the only way to proceed is by taking 'the most ambitious path.'" Bill Mollison, the father of permaculture, said that the solutions to the increasingly complex problems of the world are "embarrassingly simple." FMNR is embarrassingly simple. But I am not ashamed, because it works.

FMNR has taught me that complex, intractable problems sometimes have simple, low-cost and rapid solutions. I am hopeful. I hope that FMNR, and methods like it, will be adopted at scale and quickly in the coming years, and I continue to work towards that goal. If there is one good thing about the problems listed here, it is that they are interrelated. By tackling deforestation, we automatically tackle other pressing problems. Recent scientific research confirms that forests and other "natural climate solutions" are essential for mitigating climate change, thanks to their capacity to cool surrounding air and capture and store carbon.

Call to action

In fact, natural climate solutions can help us achieve 37%[63] of the cost-effective climate mitigation needed between now and 2030 to stabilise warming to below 2 degrees Celsius. While aggressive fossil-fuel emissions reductions are still necessary, natural climate solutions[64] offer a powerful set of options for nations to deliver on the Paris Climate Agreement while improving soil productivity, cleaning our air and water, and maintaining biodiversity. Despite this, such solutions currently receive only 2.5% of public climate financing!

"No matter where I find myself, I can assure you that FMNR is something I cannot part with for the rest of my working life." On-air in Ghana with Norbert Akolbila.

In his book, *The Weather Makers*, Australian scientist and climate action campaigner Tim Flannery says that the Keeling Curve is "one of the most wonderful things he'd ever seen, for in it you can see our planet breathing." The peaks in the saw-tooth shaped graph correspond to the northern hemisphere's winter period when little photosynthesis is occurring, and atmospheric CO_2 levels rise accordingly. The troughs in the graph correspond to spring and summer when trees of the vast boreal forests are actively photosynthesising and drawing down CO_2. I was riveted. Imagine, forests in the northern hemisphere, which are sleeping for much of the year and where photosynthesis occurs at a lower rate due to lower temperatures are having a significant impact on atmospheric CO_2 concentrations! What could a similar band of trees do in lower latitudes where temperatures are higher and growing seasons are longer?

In 2019, research conducted by ETH Zurich[66] revealed that there is enormous potential for global tree restoration—nearly one billion hectares. Importantly, the suitable sites are reported to be outside of existing forests and agricultural or urban land. Though the availability of that land, and the question of who foots the reforestation bill will need to be addressed, this news is still very good. And it only gets better! Beyond the billion hectares of land presumed to be suitable for trees, at least two billion hectares of agricultural and pastoral lands are suitable for agroforestry (trees on crop land) and silvopasture (trees on pastureland). Inclusion of trees in these already utilised areas not only makes for beautiful landscapes; it simultaneously increases their productivity and resilience while helping to mitigate climate change.

Many people in government, climate, development and agricultural circles still have a false perception that agricultural land and trees are mutually exclusive, and that globally the practice of agroforestry is of little importance. But in fact, I demonstrate in this book that with management of appropriate species, the reverse is true: trees and productive land are not only compatible, and not just complementary; for a sustainable future, they are absolutely essential. According to a study on the Global Extent and Geographical Patterns of Agroforestry,[67] the practice is already widespread; in fact, 46% of all agricultural land area globally, representing over one billion hectares of farmlands, have more than 10% tree cover. This coverage of useful trees positively influences the livelihoods of 30% of rural populations—558 million people. Additionally, tree cover on agricultural lands is increasing year by year. Of course, much good can be quickly unravelled by the short-term policies of a single political leader.

I have been able to transition from field worker to international campaigner because I am totally convinced of the effectiveness and impact of FMNR. In 35 years of promoting FMNR,

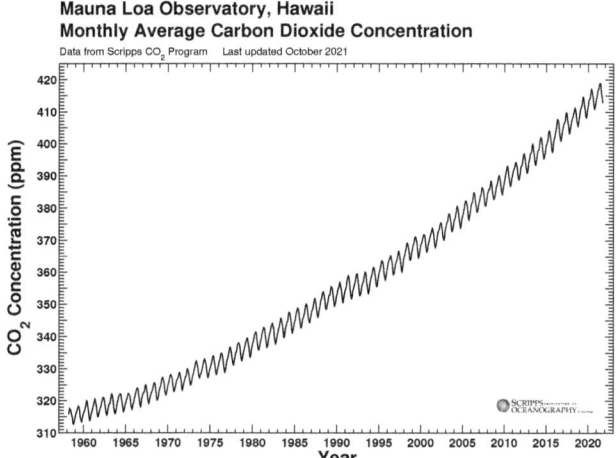

The Keeling Curve is a graph of the accumulation of carbon dioxide in the Earth's atmosphere based on continuous measurements taken at the Mauna Loa Observatory on the island of Hawaii from 1958 to the present day.[64]

nobody has ever said to me that I had ruined their land or livelihood. There has only been positive feedback. "If you have nothing to lose, and everything to gain, then by all means go for it." FMNR is like that: there are no major risks or downsides, only benefits with minimal risk. I have great confidence in giving FMNR promotion my full effort.

In my youth, I wanted God to use me to make a difference in the world. In my 20s and 30s I struggled to find a solution to deforestation and became involved in direct village interventions promoting FMNR. In my 40s my focus was on developing FMNR project proposals, attracting funding and monitoring progress. In my 50s I focused on scale-up through partnerships, face-to-face workshops and extensive travel, advocacy and promotion of FMNR. Today I ask myself, "How can I, one man, regreen the world?" And of course, the answer is, one person alone can't do it. In Niger, a small number of motivators and farmer champions showed millions of smallholder farmers how to do the impossible. On a global scale, the only way the task will be accomplished is through others—yes, governments and NGOs, yes, with donor backing—but primarily through a relatively small number of motivators and the farmers or community champions they inspire, and the millions of land users who follow their example.

Now I find myself in my 60s and have eight grandchildren. Many others currently manage FMNR projects, fundraise, design, implement, monitor and evaluate projects, and others still focus on scaling these projects up. And they do it better than I can. So, where should my focus

be now? I must help inspire the next generation and bring a diverse range of people together to work towards achieving our common goal.

Y-NOT! A call to action

Actions that prevent, halt and reverse degradation are necessary to meet the Paris Agreement target of keeping global temperature rise well below two degrees Celsius. In June 2021, the UN Decade on Ecosystem Restoration[68] called on global leaders to restore at least one billion hectares of degraded land in the next decade—an area about the size of China.

During the closing address of the 2012 Beating Famine Conference, Tony Simons, Director General of the World Agroforestry Centre, posed and answered his own question: "What is Tony spelt backwards? Y-NOT!" Some would say, that restoring one billion hectares of degraded land in ten years can't be done. I say, why not?"

If we have the capability to go to the moon, build nuclear bombs and mow down forests, then we have the capability to restore one billion hectares of degraded land and do it quickly.

If Niger, one of the poorest countries in the world, in one of the harshest climates, with minimum government or external assistance can achieve a reforestation rate of 250,000 hectares per year, what can be achieved elsewhere?

And the answer is: if we put our mind to it, much, much more.

My dream for the future is one where people and nature thrive. A place where there is hope, a bright future for children and where extreme poverty does not exist.

Will you join the movement and help restore one billion hectares of degraded land?

A miraculous result

In August 2019, during a radio interview for the Australian Broadcasting Corporation (ABC), prolific writer, filmmaker, broadcaster and provocateur, Phillip Adams said of the FMNR movement, "As a committed atheist to a committed Christian, I would have to say that these results are rather miraculous, wouldn't you?"[69]

Where there are trees there is water, food and life. Where there are trees there is hope.

Today I reflect on that boy in gumboots who yearned to plant trees on a barren hill. Feeling powerless, he reached out to a God of love and compassion and power. I see an earnest teenager and then a restless young man, with Liz by his side, who took trusting but faltering steps and who found the faith to keep moving forward despite setbacks and disappointments. More significantly, my doubts faded with the experience of seeing God at work. The God who does care, who does answer prayer and who does use very ordinary people to do extraordinary things. He is a God who has replaced despair, brokenness and tragedy with hope—and hope through, of all things, trees! The lowly tree, despised, taken for granted and abused by humankind from the dawn of time, ignored by scientists seeking technological fixes to problems of their own making, scorned by governments and industry in their blind pursuit of progress and prosperity, cut at the roots by the very farmers whose livelihoods and wellbeing depend on them—what more fitting symbol of hope than the humble and unsung tree, which freely serves humanity, in silence and forbearance and without fanfare?

In the face of environmental destruction and injustice, my journey started with a child's prayer, asking God to use me somehow, somewhere, to make a difference. I believe God honoured that prayer. I am in awe of how he has and is answering my prayer. I am grateful for the people who believed in me, supported, encouraged, taught and went before me. There is little room for doubt that God has indeed "prepared in advance good things for us to do."[70]

Act now!

 For an FMNR manual, online training, FAQs, videos, a link to donate and much more, scan the QR code and visit the FMNR Hub website.

Follow Tony on social media:

@tony.rinaudo

Tony Rinaudo

@rinaudo_tony

Acknowledgements

To all the people below and many more, you made this story possible. This is your story.

Anne Ruffer and Felix Ghezzi for spurring me on and for taking the risk of printing my story.

Suzanne Kirkbright, thank you for revealing the golden thread and keeping me to it.

Michael Collie, for drawing more of the story out of me and making it flow.

Josie Bestwick, for your sound advice and generous sharing of your family research findings.

Peter Rinaudo, for always being there for me.

The thousands of farming families and FMNR champions around the world who took my teaching and ran with it. You demonstrated that you are not passive recipients of Western aid. You proactively and enthusiastically engaged, owned and took responsibility for your own futures. Bishop Simon Chiwanga who got the "tree-stump thing" and took it unreservedly to his people. Aba Howi who dared to make dreams come true. For reasons of privacy and security some names have been changed.

SIM and EERN Associates: Ibrahim Yahaya, Moussa Housseini, Cherif Yacouba and team—you were the early pioneers who took the heat. Jim Longworth, Harry Enns, John Ockers, Emmanuel Isch, Ron Nelson, Nelson Freve, Joel Matthews, Ruth Perkins, Ed Bailey, Peter Cunningham and many, many more—you helped birth a movement. Mark Larson for proofreading proverbs and making valuable suggestions for the Hausa text.

World Vision Australia, other Support offices and World Vision International—all departments whose work enables me to do what I do, especially the many in Media, Marketing, Philanthropy, Finance and Legal. Special thanks to Tim Costello for your unstinting support, Peter Weston (early visionary). Paul Ronalds, Seak-King Huang, Mel Gow, Julianne Scenna, Tim Morris, Parin Thomson, Diarmuid Kelly, Shannon Ryan and Rob Francis ("It's a no brainer!")—for giving me space to move. Carolyn Kabore, Dean Thomson, Nick Ralph, Paul

Dettmann, Brian Hilton, Bronwyn Campbell, Cedric Hoebreck, Soheila Lew, Megan Freshwater, Jane Ogilvie, Ron Moghbelzadeh, Theresa Chew, Tori Anderson, Michelle Gale, Bruno Col, Silvio Dorati, Anne Crawford, Kate Moss, Andrew Binns and the 2030C working group, Brianna Piazza, Mike Bruce, Elissa Doherty, Susan Anderson, Graham Strong, Christopher Shore, Doug Brown, David Owens, Christophe Waffenschmidt, Dirk Bathe, Christian Cage, Richard Rumsey—and especially to our "brains trust"—the FMNR Hub past and present—Charlotte Sterrett, Sarah McKenzie, Sarah Downes, Alice Muller, Rob Kelly, you made it a team effort and fanned the flames.

Silvia Holten, FMNR champion and campaigner extraordinaire, you are in a category of your own. You gave me my voice.

World Vision National Office and Field Staff. Thank you for your over and above dedication to the cause and for putting FMNR on World Vision's and the world's radar. Special thanks to Assefa Tofu, Hailu Tafere, Kebede Regassa, Hailesellasie Desta, Caroline Njiru, Cotilda Nakyeyune, Loyce Mugisa, Mkama Nangu, Safari Dieudonne, Irene Ojouk, Festus Kiplagat, Lavenda Ondere, Lawrence Kiguro, James Ang'awa Anditi, Makabaniso Ndhlovu, Chris McMillan, Charles Bakhoum, Neil Devasahayam, Sieti Banu Emmanuel, Salmon Jacob, Moe Moe, A One, Ventia Sabathini, Amsal Ginting, Norbert Akolbila, Samuel Abasiba, Maxwell Amedi, Sumaila Seidu Sa'aka, Jesuinho Gusmao, Steve Herbert, Segenet Tessema, Goretti Goncalves Oliveira, Beatrice Mwangi, Makhera Kalele, Mafamo Pholo, Tom Roberts, Mohamed Nagueyeh Amin, Isaac Munyengingo, Michael Deng Mach and many, many more—you worked at the coal face.

Martin Falkenburg and Karsten Berends—special thanks for giving me a global platform. Right Livelihoods Foundation—for making FMNR known to the world.

Dennis Garrity, Robert Winterbottom, Chis Reij, Chris Armitage, Sebastian Matthews, Gray Tappan, Rohini Chaturvedi, Amadouu Tougiani Abas—thinkers, shakers, do-ers—you each deserve your own book. You gave FMNR global wings. Constance Neely, Mieke Bourne—peerless facilitators.

Cathy Watson, Susan Chomba, Niguse Hagazi, Joy Tukahirwa, Jonathan Muriuki, Prudence Ayebare, Pius Wamala, Alan Channer—each making your own unique contributions.

Volker Schlöndorff, through your lens, you are bringing FMNR into every home.

The individuals and churches who supported us during our time with SIM including Myrtleford, Kingston and Deepdene Uniting churches. St Alfred's Anglican Church family, our church growth group and Neville Carr, Gavin Armstrong, Boyd Eggleston, Justin Simpson—thank you for all the prayer, encouragement and coming alongside especially when things weren't working.

The thousands of donors small and large who made it possible to take FMNR to the world, including the Canadian and Australian Departments of Foreign Affairs and Trade. Special thanks to the anonymous foundation which funded the game-changing work of the FMNR Hub—you enabled World Vision to take FMNR to the world. Ulrich Bosch—beyond giving, you engaged, and you brought others on the journey. Ridley and Mieke Bell, Ruth Redpath, Amanda Jane Breidal—you saw what others didn't and made things happen.

The 141 people who donated to the Kickstarter campaign, especially Ulrich Bosch, Eleanor Lober, Milner + Huang (Alex Milner and Seak-King Huang), Joshua Prenzler and Kim and Bart Vanden Hengel.

Richard St Barbe Baker, John Steward, Roland Bunch, Alan Savory, Martin Price, Norman Vincent Peale—you led the way and inspired me.

Acknowledging people spanning many years of engagement is a dangerous exercise. If I have forgotten anybody, please take it as proof of my poor memory and not as a reflection of your own unique contribution.

Notes

1. https://iscast.org/news/tfu-garrity/
2. Volker Schlöndorff is a prominent German filmmaker. He won an Oscar as well as the Palme d'Or at the 1979 Cannes Film Festival for *The Tin Drum*, the film version of the novel by Nobel Prize-winning author Günter Grass. Schlöndorff teamed up with producer Thomas Kufus to make *The Forest Maker*, a feature-length documentary film about Tony Rinaudo and FMNR.
3. Don Watson, *The Bush: Travels in the Heart of Australia*, Melbourne, Penguin, 2014.
4. Victor Steffensen, *Fire Country: How Indigenous Fire Management Could Help Save Australia*, Sydney, Hardie Grant, 2020, p. 162.
5. *Ibid*, p. 64.
6. Ohio EPA, "The Cuyahoga 50 Years Later. Celebrating the Comeback of the Burning River," 1969–2019, film, https://youtube.com/watch?v=18JpT61rX6A
7. Richard St Barbe Baker, *I Planted Trees*, London, Lutterworth Press, 1944, p. 244
8. Matthew 7:7, NIV.
9. Matthew 7:21–23, NIV.
10. Enid and Malcolm Forsberg, *In Famine He Shall Redeem Thee: Famine Relief and Rehabilitation in Ethiopia*, Cedar Grove, SIM, 1975, pp. 117–118
11. Roland Bunch PhD is one of the most respected leaders in regenerative land management, both in terms of food security and for addressing ecological degradation and climate disruption. He has worked as a consultant in sustainable agricultural development for over 40 NGOs and governments in 50 nations, including Cornell University, the Ford Foundation, Oxfam America, Save the Children, CARE, and the governments of Guatemala, Honduras, Swaziland, Laos and Vietnam. He is the author of four books. His second book *Two Ears of Corn: A Guide to People-Centered Agricultural Improvement*, has been published in ten languages and is one of the all-time best-sellers on agricultural development programs in developing nations. Roland is an expert in green cover crops (legume crops which protect and fertilise the soil). He recognised that in more arid climates, trees play a similar role to green cover crops in adding organic matter and cooling the soil. See: www.csuchico.edu/regenerativeagriculture/international-research/roland-bunch.shtml
12. Ephesians 2:10, NIV.
13. Bruce and Norene Bond, *When Spider Webs Unite: They Can Tie Up a Lion*, self published, pp. 105–106.
14. Isaiah 43:2, NIV.
15. After 29 years in Niger, on the evening of 14 October 2016, Jeffery Woodke was abducted from his home in Abalak. Armed men killed two guards before driving Jeffery across the desert into neighbouring Mali. He has not been heard from since. See Lizzie Dearden, "American aid worker Jeffery Woodke kidnapped by suspected Islamists in Niger," *The Independent*, 15 October 2016, https://www.independent.co.uk/

news/world/africa/american-aid-worker-jeffery-woodke-kidnapped-abducted-islamists-niger-mali-aqim-alqaeda-almourabitoun-a7363771.html

16. https://en.wikipedia.org/wiki/Rangeland
17. https://en.wikipedia.org/wiki/Rainfed_agriculture
18. "Desertification," *GreenFacts*, https://greenfacts.org/en/desertification/l-3/7-climate-change-biodiversity-loss.htm
19. George Perkins Marsh, *Man and Nature: Or, Physical Geography as Modified by Human Action*, (Scribner, 1864), https://en.wikipedia.org/wiki/Man_and_Nature
20. 2 Peter 1:3, RSV.
21. Psalm 104:30, NIV.
22. Dayton Duncan, Ken Burns, *The Dust Bowl: An Illustrated History*, (San Francisco, Chronicle Books, 2012) p. 46. and Ken Burns' film documentary *The Dust Bowl*: https://kenburns.com/films/dust-bowl/
23. Samuel C. Nana-sinkam, *Land and Environmental Degradation and Desertification in Africa: The Magnitude of the Problem*, 1995. Retrieved 26 November 2007, from FAO corporate Document Repository. https://fao.org/3/X5318E/x5318e02.htm
24. Chris Reij, Senior Fellow, World Resources Institute, Washington DC. Personal communication to Tony Rinaudo, 2008.
25. Erik Eckholm, Gerald Foley, Geoffrey Barnard, and Lloyd Timberlake, *Fuelwood: The Energy Crisis that Won't Go Away*, (London, Earthscan, International Institute for Environment and Development, 1984).
26. Barry Rands, "FMNR in Niger," https://youtu.be/ZyJc3vPqOx8, and "FMNR in Niger 1990 (Part 2)," https://youtu.be/wVAZjX5rwHw
27. Joan Butterfield, Jody Butterfield and Allan Savory, *Holistic Management: A New Framework for Decision Making* (Washington DC, Island Press, 1999).
28. Sharon E. Nicholson, "The West African Sahel: A Review of Recent Studies on the Rainfall Regime and Its Interannual Variability," *International Scholarly Research Notices*, 2013. https://doi.org/10.1155/2013/453521.
29. Nathaniel Bogie, "Native Shrubs: A Simple Fix for Drought-Stricken Crops in Sub-Saharan Africa," *The Conversation*, 17 September, 2015. https://theconversation.com/native-shrubs-a-simple-fix-for-drought-stricken-crops-in-sub-saharan-africa-46125
30. When treated correctly, hanza seeds are very nutritious. Today, Sahara Sahel Foods processes and markets this and many other indigenous wild foods. See: https://uplink.weforum.org/uplink/s/uplink-contribution/a012000001pTNbqAAG/sahara-sahel-foods-using-forgotten-native-dryland-trees-as-food-sources
31. 2 Chronicles 20:15, RSV.
32. "Niger: Maternal Mortality Ratio (Modeled Estimate)," *Knoema*, https://knoema.com/atlas/Niger/topics/Health/Health-Status/Maternal-mortality-ratio
33. Katie Smith, *The Human Farm: A Tale of Changing Lives and Changing Lands* (West Hartford, Conn., Lynne Rienner, 1994).
34. Abasse Tougiani, Chaibou Guero, and Tony Rinaudo, "Community Mobilization for Improved Livelihoods through Tree Crop Management in Niger," *GeoJournal*, Volume 74, 2009, pp. 377–389.
35. Chris Reij, Senior Fellow, World Resources Institute, Washington DC. Personal communication to Tony Rinaudo, 2013.

36. "The Quesungual Agroforestry Farming System," *Food and Agriculture Organization* (FAO), https://fao.org/FOCUS/E/honduras/agro-e.htm

37. World Vision Australia (WVA) is a Christian humanitarian aid and development organisation dedicated to working with children, families and their communities to reach their full potential by tackling the root causes of poverty and injustice. World Vision serves all people, regardless of religion, race, ethnicity or gender. World Vision is an international partnership that provides short and long-term assistance to more than 100 million people worldwide in approximately 100 countries.

    The development approach of World Vision is to walk alongside vulnerable communities as they lead their own development to alleviate poverty, build resilience and tackle injustice. Child poverty is addressed in many areas of need including health, education, clean water, livelihoods and child protection. World Vision equips communities to build on their successes and, in their own time, independently continue the life-changing work we've started together.

    In 1947, American missionary Dr Robert Pierce travelled to China and Korea, where he encountered people living without food, clothing, shelter or medicine. During the Korean War in the early '50s, he helped set up orphanages to care for children who'd been abandoned or orphaned. He returned to the United States to raise funds for his work in Asia. Strong public support led to World Vision being founded in 1950, with Dr Pierce as president. World Vision Australia was established in 1969 and has since grown to become Australia's largest international non-government organisation. See: https://worldvision.com.au.

38. Lydia Polgreen, "In Niger, Trees and Crops Turn Back the Desert," *The New York Times*, 11 February, 2007, https://nytimes.com/2007/02/11/world/africa/11niger.html

39. The Gold Standard label (https://goldstandard.org) is a rigorous certification standard for carbon offset projects. It ensures their ecological and social efficacy, and delivery of genuine emission reductions and long-term sustainable development as defined by the UN Sustainable Development Goals. See: https://un.org/sustainabledevelopment/sustainable-development-goals/

40. The Clean Development Mechanism (CDM), defined in Article 12 of the Protocol, allows a country with an emission-reduction or emission-limitation commitment under the Kyoto Protocol (Annex B Party) to implement an emission-reduction project in developing countries. Such projects can earn saleable certified emission reduction (CER) credits, each equivalent to one tonne of CO_2, which can be counted towards meeting Kyoto targets. The mechanism is seen by many as a trailblazer. It is the first global, environmental investment and credit scheme of its kind, providing a standardised emissions offset instrument, CERS. A CDM project activity might involve, for example, a rural electrification project using solar panels or the installation of more energy-efficient boilers. The mechanism stimulates sustainable development and emission reductions, while giving industrialised countries some flexibility in how they meet their emission reduction or limitation targets.

41. The FMNR Hub website (https://fmnrhub.com.au) is dedicated to Farmer Managed Natural Regeneration (FMNR) with a wealth of resources and information. This website is managed by the Climate Action and Resilience team, World Vision Australia. The FMNR Hub has developed two online courses in FMNR. A short self-guided orientation and a longer,

facilitated eWorkshop. See: https://fmnrhub.com.au/fmnr-online-training

The FMNR orientation course is available for free to anyone who wants to learn about the practice of FMNR. It is self-paced and takes 2–3 hours to complete. This course focuses on: the origins of FMNR; the basics of how to do FMNR; the basics of community engagement for FMNR; the benefits of FMNR; and where to find more information on the parts of FMNR that interest participants.

The FMNR eWorkshop is a 12-week course for people who are implementing or overseeing FMNR practice in the field. It is facilitated by FMNR experts from around the world. The course includes: planning for FMNR roll-out in a community; sensitisation meetings and creating spaces for the community to discuss the impact of the environment on their well-being; understanding the physical FMNR practice and tips for teaching it; identifying and selecting individuals who will become FMNR champions; building relationships in communities and ensuring FMNR is inclusive; troubleshooting when FMNR isn't going well; and where to find more FMNR information.

For any questions please email: livelihoods_wvecampus@wvi.org

42. Online training has proven to be a very effective tool for building the capacity of field staff and project managers. It has enabled far greater numbers of practitioners at lower cost than face-to-face training. Participants have come through the course not only with a strong grounding in introducing and promoting FMNR, but also in how to engage with rural communities and introduce change. Since its inception in 2015 some 200 participants from 28 countries have been trained. See: https://fmnrhub.com.au/fmnr-online-training

43. The 190-page *FMNR Manual* distils over 30 years of experience and best-practice learning in FMNR. Chapters cover a broad range of topics including: the history and definition of FMNR; community engagement; how to overcome obstacles; and the design, monitoring and evaluation of FMNR projects. The manual is available in English and French at: https://fmnrhub.com.au/wp-content/uploads/2019/03/FMNR-Field-Manual_DIGITAL_FA.pdf

44. Beating Famine Conference, Nairobi, 10–13 April 2012. https://beatingfamine.com

45. "What is Evergreen Agriculture?" *Evergreening Agriculture*, https://evergreenagriculture.net/what-is-evergreen-agriculture

46. FMNR has been featured in distinguished outlets such as German weekly *Der Spiegel*, the Australian Broadcasting Corporation's *Lateline* (TV channel), ZDF TV *News* (Germany), *The Guardian* (a UK daily newspaper), PBS *NewsHour* (US public service TV channel), *The Smithsonian* and *The New Yorker*.

47. Malcolm Gladwell, *The Tipping Point: How Little Things Can Make a Big Difference*, New York, Little Brown, 2000.

48. Liz Rinaudo, "Releasing the Underground Forest in Mpwapwa, Tanzania," *FMNR Hub*, 22 September, 2012, https://fmnrhub.com.au/releasing-the-underground-forest-in-mpwapwa-tanzania/

49. https://leadfoundation.org

50. https://awakentrees.org

51. From 2012, the World Agroforestry Centre (ICRAF) began researching and championing FMNR, and actively seeking partnerships with World Vision and other NGOs. The United States Agency for International Development (USAID) has listed FMNR as a best-practice intervention for Natural

Resource Management (NRM) and resilience projects. The World Bank and others, in the quest for more "climate-smart agriculture," have also recognised the important role of agroforestry noting the success of FMNR and EverGreen Agriculture in Niger (The World Bank, Climate Smart Agriculture. See: https://worldbank.org/content/dam/Worldbank/document/CSA_Brochure_web_WB.pdf

In 2020 UN experts included FMNR amongst 512 good practices which are now featured on their dedicated website (sustainabledevelopment.un.org/partnership/?p=30735). This is what they wrote about FMNR: "We are pleased to inform you that your initiative has been selected to be included in a digital publication which will feature selected outstanding Sustainable Development Goal (SDG) Good Practices. It is expected that these practices will continue to inspire governments and stakeholders in their effort to accelerate implementation of the 2030 Agenda at all levels.

52. Roland Bunch, during a presentation to a national FMNR conference in Addis Ababa following the 2012 Beating Famine Conference in Nairobi.

53. The Global Evergreening Alliance (GEA) provides a collaborative platform to support and facilitate massive-scale environmental restoration and sustainable agricultural intensification projects at a globally significant scale. The Alliance works with, and through, a multitude of member organisations, institutions and governments at the national and subnational levels across Sub-Saharan Africa, Latin America and Southeast and South Asia, empowering millions of farmer families across the world to restore their land, lives and livelihoods, to scale up nature-based solutions to restore degraded lands and capture carbon, with a goal of capturing 20 billion tonnes of CO_2 annually by 2050. One of the major interventions of the Alliance is the promotion of FMNR. In the words of GEA Chairman Dennis Garrity's words: "We do not need a moonshot, hit-and-miss approach to keeping the earth habitable. We know what to do and it is within our means to do it. All we need is the will." https://evergreening.org

54. https://greatgreenwall.org

55. Horand Knaup, "Mister Rinaudo will die Wüste stoppen," *Der Spiegel*, 18 June 2012.

56. https://rightlivelihoodaward.org

57. Michael Williams, *Deforesting the Earth: From Prehistory to Global Crisis* (Chicago: University of Chicago Press, 2003).

58. "World population projected to reach 9.8 billion in 2050, and 11.2 billion in 2100," *United Nations*, 21 June, 2017, https://un.org/development/desa/en/news/population/world-population-prospects-2017.html, and "World Population from 1950 to 2020," *Statista*, 24 August, 2021, https://de.statista.com/statistik/daten/studie/1716/umfrage/entwicklung-der-weltbevoelkerung

59. *Soil Atlas, Facts and Figures about Earth, Land and Fields*, (Berlin: Heinrich Böll Foundation, and Potsdam: Institute for Advanced Sustainability Studies, 2015), https://boell.de/soilatlas

60. See NYDF 2019 Progress Report: "Protecting and Restoring Forests: A Story of Large Commitments yet Limited Progress," https://climatefocus.com/sites/default/files/2019NYDFReport.pdf, and https://forestdeclaration.org

61. See IBPES Report 2019. *Nature's Dangerous Decline "Unprecedented" Species Extinction Rates "Accelerating"* 2019. https://un.org/sustainabledevelopment/blog/2019/05/nature-decline-unprecedented-report/

62. The UN Secretary General called the latest IPCC climate report "code red for humanity," stressing

irrefutable evidence of human influence. See: Nils Rokke, "COP26: What Must Happen After Code Red on Climate Change," *Forbes*, 28 October, 2021, https://forbes.com/sites/nilsrokke/2021/10/28/cop26-what-must-happen-after-code-red-on-climate-change/

63. Bronson W. Griscom, *et al.*, "Natural Climate Solutions," *Proceedings of the National Academies of Science of the United States of America*, 31 October 2017, https://pnas.org/content/114/44/11645

64. http://naturalclimatesolutions.org

65. https://keelingcurve.ucsd.edu/
https://scrippsco2.ucsd.edu/

66. Jean-Francois Bastin *et al.*, "The global tree restoration potential," *Science*, 365, 2019, 76–79. See also: "Nature-Based Solutions in the Fight against Climate Change", TEDx Talks, 10 December 2019, https://youtube.com/watch?v=CSH63qgpG0Y

67. Robert J. Zomer *et al.*, "Global Extent and Geographical Patterns of Agroforestry," *The Overstory*, 14 February, 2010.

68. World Vision is an official supporting partner with this initiative because of their work in promoting FMNR. See: https://decadeonrestoration.org

69. https://www.abc.net.au/radionational/programs/latenightlive/the-forest-maker/11450332

70. "For we are his workmanship, created in Christ Jesus to do good works, which God prepared in advance for us to do." (Ephesians 2:10, NIV)

Photo credits

Silas Koch World Vision	Front cover, 166, 171, 175, 176
Mouslim Sidi Mohamed World Vision	8
lkonya Alamy Stock Photo	12
Tony Rinaudo collection	14, 15, 20, 25, 26, 28, 29, 33, 36, 38, 44, 50, 53, 59, 60, 76, 80, 82, 94, 103, 106, 114, 116, 120, 122, 123, 124, 127, 128, 131, 134, 138, 154, 159
State Library of Victoria	16, 18, 19
Myrtleford Times	22
Myrtleford and District Historical Society	24
Roberto Rossellini Berit Films	27
Cleveland Public Library	30
University of Saskatchewan University Archives and Special Collections	34
Ian Anderson University of Saskatchewan University Archives and Special Collections	35
Steve Reynolds World Vision	42
Roland Bunch Sustainable Harvest International	47
Jon Warren	56, 64, 66, 68, 72, 74, 78, 84, 92
Johannes Dieterich	90, 148, 162, 168, 198

Tony Rinaudo World Vision	86, 96, 101, 110, 118, 143, 146, 152, 159, 178, 178, 182, 185, 186, 188, 190, 192, 204
Arthur Rothstein Library of Congress	103
Rüffer & Rub	109
World Vision	112, 160, 180, 194, 201
Ross Hoddinott Alamy Stock Photo	157
Paul Bestwick	164
Gray Tappan US Geological Survey	173
Wolfgang Schmidt Right Livelihood Award Foundation	195
Suzy Sainvoski World Vision	Back cover

www.ingramcontent.com/pod-product-compliance
Lightning Source LLC
Chambersburg PA
CBHW061805290426
44109CB00031B/2939